The Quantum Tapestry: Unraveling Non-Linear Time and the Mandela Effect

KATHLENE HERBERGER

Table of Contents

DEDICATION

To the tireless explorers of consciousness, those who dare to question the accepted narratives and seek truth beyond the confines of conventional wisdom. This work is dedicated to the unsung heroes of intellectual curiosity, whose relentless pursuit of knowledge expands the boundaries of human understanding. It is also dedicated to those who have experienced the uncanny – the déjà vu, the precognitive dreams, the seemingly impossible synchronicities – for it is through your experiences that the whispers of non-linear time become audible. May this book serve as a testament to the courage it takes to confront the mysteries that lie at the edge of perception, and may it inspire future generations to continue this vital exploration of the enigmatic nature of time. Finally, this work is dedicated to the memory of those who have passed, their journeys interwoven with threads of time that may stretch far beyond our current comprehension. Their legacies are a constant reminder of the enduring power of consciousness and its profound connection to the unfolding fabric of reality. Their lives, though finite in linear time, resonate across a potentially boundless temporal landscape

ACKNOWLEDGMENTS

I am deeply grateful for the intellectual stimulation and thought-provoking conversations that have shaped this work. Finally, my heartfelt thanks go to my family and friends for their endless patience and understanding throughout this long and often challenging journey. Their unwavering love and support have been my constant source of strength.

1 Introduction to Non-Linear Time

Our understanding of reality is fundamentally shaped by our perception of time. For most of human history, and for the vast majority of individuals today, that perception is inherently linear. We experience time as a unidirectional flow, a relentless march from a fixed past, through a fleeting present, towards an uncertain future. This linear model, deeply ingrained in our language, culture, and even our neurological structures, dictates our understanding of causality, free will, and the very nature of existence. We assume cause precedes effect, that events unfold in a predictable sequence, and that our actions are confined to the present moment, influencing only the future. But is this linear conception of time an accurate reflection of reality, or is it merely a convenient, albeit limiting, framework for navigating our experience?

The linear model of time is so pervasive that we often take it for granted, failing to question its underlying assumptions. We organize our lives around it, scheduling appointments, setting deadlines, and planning for the future based on the premise of a straight-line progression. Our legal systems, our historical narratives, and our personal memories all operate within this linear framework. Yet, a growing body of evidence, drawn from diverse fields such as quantum physics, parapsychology, and even anecdotal experiences, suggests that our linear perception may be, at best, an incomplete, and at worst, a fundamentally flawed representation of how time actually functions.

Psychological studies on temporal perception offer valuable insights into how we experience time, but they often fall short of providing a complete picture. Experiments consistently demonstrate the subjective nature of time, showing how our perception of duration is influenced by factors such as attention, emotional arousal, and the presence of significant events. When we are bored, time seems to drag; when we are engrossed in an exciting activity, it flies by. These findings highlight the plasticity of our temporal experience, indicating that our internal clock isn't a precise measure of objective time, but rather a subjective construction influenced by a complex interplay of cognitive and emotional factors.

However, these studies primarily focus on variations in the *perception* of linear time, not the possibility of time itself being fundamentally non-linear. They account for discrepancies by attributing them to cognitive biases or neurological processes, effectively remaining within the confines of the linear model. They don't address phenomena that seem to defy linear causality, such as precognitive dreams, déjà vu experiences, or the seemingly inexplicable synchronicities that occasionally punctuate our lives. These anomalies, often dismissed as coincidences or attributed to psychological quirks, challenge the very foundation of our linear understanding of time and suggest the existence of a more complex, perhaps even non-linear, temporal reality.

The limitations of psychological studies become even more apparent when we consider the persistent reports of shared false memories, a phenomenon widely known as the Mandela Effect. Countless individuals vividly recall events that never actually happened, sharing these collective misrememberings with striking consistency. While psychological explanations, such as memory distortion or confabulation, are frequently offered, these explanations fail to fully account for the widespread and persistent nature of these shared illusions. The Mandela Effect, therefore, presents a compelling challenge to our understanding of memory, and by extension, our understanding of time itself. If our collective memories can be so profoundly altered, so fundamentally at odds with objective reality, then perhaps our perception of the linear flow of time itself is also susceptible to distortions that transcend the purely psychological realm.

The challenge to the linear model doesn't originate solely from anecdotal evidence or seemingly anomalous psychological phenomena. The field of quantum mechanics, despite its apparent strangeness, provides an even more profound challenge to the classical, linear understanding of time. The Heisenberg Uncertainty Principle, for example, implies that we cannot simultaneously know both the position and momentum of a particle with perfect accuracy. This inherent uncertainty at the quantum level calls into question the deterministic nature of linear time, suggesting that the future isn't predetermined, but rather exists as a probability cloud of potential outcomes.

Quantum entanglement further complicates the picture. This phenomenon involves two or more particles becoming linked in such a way that they share the same fate, regardless of the distance separating them. Measurements performed on one entangled particle instantaneously affect the state of the other, even if they are light-years apart. This "spooky action at a distance," as Einstein famously called it, suggests a connection that transcends the limitations of space and time as we traditionally understand them, hinting at a deeper interconnectedness that challenges the boundaries of linear causality.

The implications of quantum mechanics extend beyond mere uncertainty and non-locality. Some interpretations of quantum theory, such as the Many-Worlds Interpretation, posit the existence of parallel universes, each representing a different possible outcome of a quantum measurement. In this framework, time isn't a single, linear stream, but rather a branching, multi-dimensional structure, where every decision leads to a new reality, a new timeline. This radical departure from the linear model opens up the possibility of alternate pasts and futures, existing simultaneously but independently of our own perceived reality. While still largely theoretical, the Many-Worlds Interpretation offers a compelling framework for understanding phenomena that defy linear causality, and for conceptualizing a reality far richer and more complex than our everyday experience suggests.

Beyond quantum mechanics, other scientific theories also contribute to the burgeoning critique of linear time. Einstein's theories of relativity, for instance, demonstrate the relativity of simultaneity, showing that the perception of whether two events occur simultaneously is dependent on the observer's frame of reference. General relativity, further, suggests that time is not absolute, but rather is interwoven with space, forming a four-dimensional spacetime continuum that can be warped and distorted by gravity. These theories demonstrate that time is not the immutable, objective entity we often assume it to be; it's a flexible, relativistic concept whose properties are influenced by the gravitational fields and the relative velocities of observers. This relativistic view of time, while not explicitly endorsing non-linearity, at least opens the door to the possibility that the linear model may be an insufficient description of time's true nature.

The linear model of time, despite its ubiquity, is a construct, a simplification that serves our practical needs but may not accurately reflect the underlying reality. The anomalies that challenge this model – from the

Mandela Effect to quantum entanglement, from déjà vu to precognitive dreams – point towards a more nuanced, perhaps even non-linear, understanding of time. The exploration of non-linear temporal theories is not a rejection of scientific rigor; it's a necessary expansion of our framework, a broadening of our perspective to encompass the complexities of the universe and our experience within it. The journey into non-linear time is an exploration of the very fabric of reality, a quest for a deeper understanding of ourselves and the cosmos we inhabit. The following chapters will delve deeper into the specific models of non-linear time, exploring their implications for our understanding of causality, free will, and the nature of reality itself.

The preceding discussion established the limitations of a purely linear model of time, highlighting its inadequacy in explaining certain phenomena. Now, we delve deeper into the evidence that directly contradicts the linear paradigm, focusing on anecdotal and scientific observations that point towards a more complex, and potentially non-linear, temporal reality.

One of the most compelling pieces of anecdotal evidence comes from the realm of personal experience. Déjà vu, that unsettling feeling of having already lived a current moment, defies linear causality. The sensation of familiarity, often accompanied by a sense of unease, is jarring precisely because it violates our expectations of linear temporal progression. Within a linear framework, an experience can only be familiar if it has already occurred in the past. Déjà vu, however, presents the inexplicable feeling of familiarity with an experience happening *for the first time*. Conventional explanations frequently cite neurological glitches or memory misattributions, suggesting a malfunction in the brain's processing of temporal information. However, the widespread prevalence of déjà vu, and the deeply ingrained feeling of genuine familiarity it evokes, suggests something more profound than mere neural noise. Could it be a glimpse into a parallel timeline, a fleeting resonance with a past or future self, or a manifestation of a non-linear temporal interaction? While these interpretations remain speculative, they offer a framework for understanding déjà vu beyond the confines of conventional neurological explanations.

Similar to déjà vu, precognitive dreams also challenge the linear model. These are dreams that seemingly predict future events, events that the dreamer could not have reasonably known about beforehand. Numerous accounts exist, ranging from seemingly trivial details to life-altering predictions, suggesting a potential connection between dreams and future events that transcends the limits of linear causality. While skepticism surrounding precognitive dreams is understandable, the sheer volume of such reports, and the striking accuracy of some predictions, warrant serious consideration. Dismissing them outright as coincidence ignores the possibility of a more complex interplay between consciousness, time, and reality. The study of precognitive dreams, often relegated to the fringes of scientific inquiry, could unlock critical insights into the nature of time if approached with an open mind and rigorous methodology. Furthermore, the exploration of these phenomena requires moving beyond simply labeling these experiences as "coincidence" and undertaking the challenging, but necessary, task of devising rigorous testing protocols to eliminate biases and allow for systematic investigation.

The inconsistencies often discovered within our personal memories also challenge the linear perception of time. Our memories are not simply accurate recordings of past events, but rather reconstructions shaped by our emotions, biases, and the passage of time. These reconstructions can be influenced by our current state, by later experiences, or even by suggestion. However, certain inconsistencies go beyond simple memory distortions. The Mandela Effect, as discussed previously, presents a particularly intriguing challenge. The widespread sharing of false memories—the recollection of events that never occurred—suggests a collective distortion of temporal reality. While neurological explanations attempt to account for individual cases of memory alteration, they fall short of explaining the shared nature of the Mandela Effect. This suggests a possible influence that transcends the individual consciousness, perhaps even interacting with a broader, collective memory field that may not adhere to the constraints of linear time. Further research into the neurobiological mechanisms behind memory formation and retrieval, combined with a comparative analysis across multiple documented cases of the Mandela Effect, could potentially reveal underlying patterns or structures that might help us understand this shared distortion of past experiences. Exploring this area rigorously can enhance our understanding of the boundaries between individual memory and a possible collective memory matrix.

Furthermore, the study of synchronicities, seemingly meaningful coincidences, also throws the linearity of time into question. These are events that appear connected in a way that defies chance, often exhibiting a striking relevance to the observer's life or current situation. While dismissals are easy, such events often carry a profound emotional impact, suggesting a deeper significance beyond random chance. Carl Jung's concept of synchronicity, though often criticized for its lack of empirical evidence, proposes a framework for understanding these events as meaningful coincidences that point to a deeper interconnectedness within the universe, a connection potentially mediated by non-linear temporal dynamics. To objectively study synchronicities, it would be beneficial to develop a more sophisticated metric for quantifying the probability of these events occurring purely by chance.

Beyond anecdotal evidence, scientific theories also directly challenge the linear model. The quantum realm, as mentioned earlier, presents a multitude of phenomena that seem to defy the rules of classical physics and the accompanying linear perception of time. Quantum entanglement, with its non-local correlations, demonstrates instantaneous interactions between particles regardless of their spatial separation. This "spooky action at a distance" implies a connection that transcends the limitations of space and time as we conventionally understand them, suggesting an underlying reality where information or influence might propagate faster than the speed of light or even outside the limitations of space and time as we currently grasp them. Exploring this interaction in the context of time requires formulating new theoretical models that account for non-locality within a temporal framework that transcends the linear.

Quantum tunneling, another quantum phenomenon, shows that particles can seemingly pass through energy barriers, defying classical mechanics and hinting at a possible non-linearity in temporal progression. This process suggests that particles might not simply progress linearly through time, but can, under specific conditions, "jump" across different points in time or even different energy states in a non-linear manner.

Moreover, the various interpretations of quantum mechanics themselves provide alternative models of time. The Many-Worlds Interpretation, for example, proposes a branching structure of timelines where every quantum measurement creates a new universe, thus creating a multiverse where potentially every decision leads to a new reality. This radically departs from the singular, linear progression of time and suggests that time may not be singular, but a complex network of branching possibilities. This perspective opens a wide array of avenues for exploring the subjective nature of experience within a non-linear temporal framework.

The implications are staggering. If time is not linear, our understanding of causality, free will, and our place in the universe must be fundamentally revised. The exploration of non-linear temporal theories is not merely a speculative exercise, but a necessary step towards a more complete and accurate understanding of reality. The evidence presented, while still open to interpretation and further investigation, points towards a model of time far richer and more complex than the simple linear model that has dominated our thinking for so long. The next chapters will continue this exploration, introducing specific models of non-linear time and examining their implications for various aspects of human experience and the universe at large.

Having established the limitations of a strictly linear temporal model and explored the anomalous phenomena that challenge its validity, we now turn to alternative frameworks that offer potentially more comprehensive explanations. These models move beyond the simplistic "arrow of time" and embrace the complexities suggested by quantum mechanics, anecdotal evidence, and cultural narratives. Let us begin by examining the concept of cyclical time.

Unlike the linear model, which envisions time as a unidirectional progression from past to future, cyclical time proposes a recurring pattern. Imagine a wheel constantly rotating; the past, present, and future are not distinct points but parts of a continuous cycle. This model resonates with many ancient cultures, who saw time as a repeating series of ages or epochs, mirroring the cycles of nature – the seasons, the tides, the phases of the moon. Ancient Egyptian cosmology, for instance, depicted time as a cyclical process of creation, destruction, and rebirth, mirroring the cyclical flooding of the Nile River which was essential to their agricultural society. Similarly, many indigenous cultures maintained cyclical calendars, reflecting their understanding of time's recurrent nature. The Mayan calendar, with its intricate cycles and predictions, is a prime example, albeit one often misinterpreted in popular culture. These cyclical models, while seemingly simple, offer a framework for understanding events that seem out of place within a strictly linear paradigm. The repetition inherent in cyclical time could account for déjà vu experiences, as a recurrence of a previously lived moment within the cycle. It could also potentially offer a different lens through which to view the Mandela Effect. Instead of a collective memory distortion, it could represent a slippage or bleed-through between different iterations within the cyclical pattern. A particular "epoch" in the cyclical model might share similarities with another, causing a blend of memories or perceptions.

However, the purely cyclical model, while offering a compelling alternative to linearity, may also present its own limitations. It struggles to account for the perceived irreversibility of certain processes, such as entropy, which generally increases over time. Furthermore, the strict periodicity implied by a simple cyclical model might be an oversimplification. It doesn't readily accommodate the apparent uniqueness and progression observed in many evolutionary and historical processes. A more nuanced approach might involve considering nested cycles, with smaller cycles embedded within larger ones, allowing for both recurrence and overall progression. This could be envisioned like the rotation of planetary bodies within a solar system – the smaller cycles of daily rotations and monthly orbits are part of a larger, longer cyclical movement around the galaxy.

Another model that significantly departs from linear time involves the concept of parallel universes or branching timelines. This model, often associated with the Many-Worlds Interpretation of quantum mechanics, proposes the existence of multiple universes, each representing a different outcome of a quantum event. Imagine a tree with its trunk representing the present moment and its branches representing different possible futures. Every decision, every quantum event, generates a new branch, leading to a separate universe where that choice has played out differently. In this context, phenomena like precognitive dreams could be interpreted as glimpses into parallel timelines, where the dream event has already occurred in one of these alternate realities. Déjà vu, within this model, could represent a brief overlap or resonance between our primary timeline and a parallel one, creating a feeling of familiarity with an experience that, from our immediate perspective, is novel. The implications of this model are far-reaching, suggesting the existence of countless versions of ourselves, living out countless variations on our lives. The Mandela Effect, too, finds a potential explanation within this multiverse hypothesis. The shared false memory could represent a collective "bleed-through" from a parallel universe where the remembered event did in fact occur. While the precise mechanisms of such interactions remain speculative, the multiverse hypothesis provides a framework for considering these paradoxical events without the need to reject our sensory experiences or our understanding of consciousness.

Closely related to the parallel universes concept is the notion of time loops. Time loops posit that time doesn't simply proceed linearly but can fold back on itself, creating repeating cycles of events. This is a common theme in science fiction, often portrayed as a self-contained loop where individuals are trapped in a repeating time period until they solve a specific problem or break the loop. This model, while seemingly fantastical, offers another possible explanation for some of the perplexing phenomena we've discussed. Events that appear to be coincidences or instances of precognition could be interpreted as segments within a larger time loop. Deja vu might represent the conscious recognition of a previously experienced moment within the loop. The apparent "randomness" or "meaningfulness" of synchronicities could be interpreted as the self-consistent structure of a closed time loop designed by a complex set of cause-and-effect relationships that we are only beginning to grasp. While empirical evidence for time loops remains elusive, their theoretical exploration offers an alternative model that challenges the very structure of linear causality. One significant challenge in testing the hypothesis of time loops lies in the potential self-referential paradox created. Observing or measuring a time loop would inherently alter it, making objective observation inherently impossible, at least from within the loop itself.

In summary, moving beyond the limitations of linear time necessitates exploring alternative models that incorporate the complexities suggested by quantum phenomena, anecdotal evidence, and cross-cultural perspectives. Cyclical time, parallel universes, and time loops represent three distinct yet interconnected frameworks that offer potential explanations for various puzzling events. The inherent limitations of each model highlight the need for a more integrative approach. The next section will delve further into the integration of these models, exploring the potential for a composite model of non-linear time that accommodates the strengths and addresses the shortcomings of each individual approach. This pursuit of a unified understanding will ultimately aim at creating a more complete and accurate representation of temporal reality, one that acknowledges the intricate interplay between subjective experience, collective consciousness, and the fundamental laws governing our universe. This necessitates a shift away from rigid, reductionist explanations and towards a more holistic, open-minded approach to the exploration of time itself. The exploration of non-linear time requires a willingness to embrace complexity, acknowledge limitations in current scientific understanding, and consider perspectives outside the mainstream scientific paradigm.

Our journey into the realm of non-linear time now necessitates a deeper exploration of quantum mechanics, a field that fundamentally challenges the deterministic worldview underlying the classical, linear conception of time. The counterintuitive nature of quantum mechanics, often described as "spooky action at a distance" by Einstein himself, offers a fertile ground for rethinking our understanding of temporality. Classical physics, with its predictable trajectories and causal chains, provides a comfortable, albeit limited, framework for comprehending the world. However, the microscopic world unveiled by quantum mechanics reveals a reality far stranger and more nuanced.

At the heart of this strangeness lies the Heisenberg Uncertainty Principle, a cornerstone of quantum theory. This principle states that it is fundamentally impossible to simultaneously know both the position and momentum of a particle with perfect accuracy. The more precisely we determine one, the less precisely we can determine the other. This inherent uncertainty isn't simply a limitation of our measuring instruments; it's a fundamental property of the universe at the quantum level. This implies an inherent indeterminism; the future trajectory of a quantum particle is not predetermined, but rather exists as a probabilistic superposition of possibilities. This probabilistic nature profoundly impacts our understanding of causality. In a classical, deterministic universe, every effect has a clearly defined cause. In the quantum realm, however, causes and effects become blurred, intertwined in a probabilistic dance.

Consider, for instance, the famous double-slit experiment. When electrons are fired at a screen with two slits, they don't simply pass through one slit or the other as classical particles would. Instead, they exhibit wave-like behavior, creating an interference pattern on the screen. This suggests that the electron exists in a superposition of states, passing through both slits simultaneously until the moment of measurement. This act of observation, the very process of trying to determine its position, collapses the superposition, forcing the electron to "choose" a specific path. The implication is that the electron's path wasn't predetermined until the moment of measurement; the act of observation itself influences the outcome. This challenges the notion of a pre-ordained, linear progression of events. The electron's past and future are not fixed until the moment of observation, suggesting a temporal fluidity that contradicts the classical framework.

Furthermore, the concept of quantum entanglement throws another wrench into our linear understanding of time. Entanglement refers to a phenomenon where two or more quantum particles become linked, regardless of the distance separating them. Changes made to one entangled particle instantaneously affect the other, even if they are light-years apart. This instantaneous correlation presents a profound challenge to the concept of locality, the idea that an event can only influence its immediate surroundings. Entanglement implies a deeper interconnectedness, a non-local correlation that transcends the limitations of space and, potentially, time. The influence seems to be instantaneous, seemingly violating the constraint of the speed of light, which is the universal speed limit for information transfer according to Einstein's theory of special relativity. Some interpretations suggest that entangled particles share a common past, present, and future, regardless of spatial separation, implying a non-linear interconnectedness across time.

The implications of these quantum phenomena for our understanding of time are far-reaching. The indeterministic nature of quantum mechanics, coupled with non-locality, suggests that time might not be the unidirectional, immutable flow envisioned in classical physics. Instead, time could be a more fluid, flexible dimension, influenced by the act of observation and the interconnectedness of quantum systems. The probabilistic nature of quantum events suggests that multiple possibilities exist simultaneously, until a specific outcome is observed. This resonates with the multiverse hypothesis mentioned earlier, where each quantum measurement creates a branching timeline, resulting in a vast ensemble of parallel universes. In this context, our subjective experience of linear time could be merely a limited perspective within a much larger, multidimensional temporal reality.

The challenge, then, is to reconcile these quantum insights with our everyday experience of linear time. While the quantum world governs the subatomic realm, our macroscopic world appears to follow relatively linear trajectories. The emergence of classical behavior from quantum mechanics remains one of the most significant unsolved problems in physics, known as the "measurement problem." However, the very existence of this problem highlights the limitations of our current understanding. It is possible that our perception of linear time is an emergent property of macroscopic systems, an approximation that breaks down at the quantum level.

Furthermore, the implications of quantum mechanics extend beyond physics and into the realms of consciousness and information theory. Some researchers propose that consciousness itself might play a crucial role in the collapse of the quantum wave function, suggesting a fundamental link between consciousness and the nature of time. This link is explored in various interpretations of quantum mechanics, such as the von Neumann-Wigner interpretation, which posits that consciousness is necessary for the wave function to collapse. The implications of such a view are profound, suggesting that the observer, the conscious being, is not simply a passive recipient of information but an active participant in shaping reality, including the flow of time.

In the context of non-linear time, these perspectives offer intriguing possibilities. If consciousness is intrinsically linked to the collapse of the wave function, then our subjective experience of time could be influenced by our conscious choices and observations. Precognitive experiences, for instance, could be interpreted as glimpses into alternative quantum pathways, possibilities that exist in superposition until the moment of observation, when one path is selected. Similarly, déjà vu might reflect a momentary overlap or resonance with another quantum state, creating a fleeting sense of familiarity with an experience that has not yet manifested in our primary timeline.

The investigation into the intersection of quantum mechanics and time continues to be a vibrant and controversial field. The intricacies of quantum theory challenge our fundamental assumptions about causality, locality, and even the nature of reality itself. The implications for our understanding of time are profound, hinting at a reality far richer and more complex than the simplistic linear model. As we delve deeper into the quantum realm, we find ourselves venturing into uncharted territory, forcing us to rethink our most basic assumptions about the nature of time, space, and consciousness. The journey is far from over, yet the path ahead is filled with both fascinating possibilities and intriguing challenges. The inherent uncertainties of the quantum world necessitate a move away from rigid, deterministic models towards a more open-ended, explorative approach that embraces the complexity and mystery at the heart of reality. The next section will explore how these quantum considerations can inform our understanding of specific phenomena typically associated with non-linear time, such as precognition, déjà vu, and the Mandela Effect, bringing us closer to a more comprehensive model of temporal reality.

The exploration of non-linear time thus far has led us to the fascinating and often perplexing world of quantum mechanics. Its inherent probabilistic nature, non-locality exhibited through entanglement, and the profound implications of the observer effect all challenge our conventional understanding of a strictly linear progression of time. These principles, however counterintuitive they may seem within our everyday macroscopic experience, open doors to a broader conceptualization of temporality, one that embraces the possibility of multiple realities, overlapping timelines, and the influence of consciousness on the unfolding of events. This brings us to a specific phenomenon that exemplifies the potential intersection of quantum mechanics and non-linear time: the Mandela Effect.

The Mandela Effect, named after the widespread false memory that Nelson Mandela died in prison during the 1980s, is a prime example of a shared false memory. It's not merely an isolated incident of individual misremembering; it represents a collective misrecollection, experienced by numerous individuals independently. The scale of these shared illusions is significant enough to warrant serious consideration, particularly in light of our discussion of quantum mechanics and the potential for multiple realities or overlapping timelines. Thousands, if not millions, of people genuinely believe they remember details of events or products that never actually occurred as they remember them. This isn't simply a matter of faulty memory; the consistency and widespread nature of these shared false memories defy easy explanation within traditional psychological frameworks.

Examples abound. Many people vividly recall the children's book featuring a misspelling of the iconic phrase "Luke, I am your father" as "Luke, I am your father" (with the missing 's'). Others swear that the Star Wars character, Darth Vader, says "No, I am your father," which is indeed the correct and actual dialogue. This slight yet significant variation perfectly demonstrates the strange nature of the Mandela Effect. Numerous people remember the logo of the famous candy company "Fruit of the Loom" featuring a fruit bowl, yet the logo always has only featured a cornucopia. Similarly, the spelling of "Jiffy" peanut butter is frequently misremembered as "Jiffy", with various other spelling deviations. This consistent pattern across various brands and pop culture elements highlights the scale of the issue. The consistency in misremembering, with vast numbers of people reporting identical distortions, strongly suggests something more complex than mere individual memory lapses.

The psychological explanations offered for the Mandela Effect often focus on memory distortions, confabulation, and the influence of social contagion. Cognitive biases, such as confirmation bias and the availability heuristic, might contribute to the reinforcement and spread of these false memories. Individuals might initially misremember a detail, and through subsequent discussions and online interactions, this misinformation becomes amplified and solidified, creating a self-reinforcing loop of shared misremembering. The power of suggestion and the inherent malleability of memory are undeniable factors.

However, the sheer scale and consistency of these shared false memories, extending beyond isolated incidents to encompass numerous examples across diverse cultural contexts, pose a challenge to purely psychological explanations. The prevalence of such widespread and identical false memories across vast geographic distances and demographic groups is difficult to fully explain using conventional memory models alone. While social contagion undeniably plays a role in spreading these false memories, it doesn't entirely explain their initial appearance. It doesn't explain why so many individuals independently arrive at the same inaccurate recollection.

This is where the intriguing possibilities arising from non-linear temporal theories come into play. Could the Mandela Effect be a manifestation of parallel realities or overlapping timelines, where individuals momentarily glimpse or experience fragments of alternate realities, resulting in these shared false memories? This perspective moves beyond a purely psychological interpretation and suggests a possible connection to the deeper, more fundamental levels of reality touched upon by quantum mechanics. If, as some theories propose, there exist multiple versions of reality or alternate timelines, a temporary overlap or 'bleed-through' could explain these shared illusions. This overlap might be facilitated by specific quantum phenomena or consciousness-related processes, as previously discussed in relation to the collapse of the wave function.

Within the framework of a multiverse hypothesis, numerous parallel universes exist simultaneously. Each quantum event might generate branching timelines, creating a vast ensemble of parallel realities. The Mandela Effect could then be interpreted as a fleeting glimpse into one of these alternative timelines, a temporary interaction between our perceived reality and another one. This interaction, even if transient, would leave an imprint on the collective consciousness, manifesting as the shared false memories observed in the Mandela Effect.

This interpretation, however radical, aligns with some interpretations of quantum mechanics that posit a participatory role for consciousness in shaping reality. If consciousness is not simply a passive observer but an active participant in the collapse of the wave function, then it could influence the selection of a particular quantum pathway, thereby affecting the unfolding of events within a given timeline. The Mandela Effect might represent moments where the boundaries between these timelines become temporarily permeable, allowing for these glimpses into alternative realities.

The debate surrounding the Mandela Effect is far from settled. Some researchers maintain that psychological explanations alone are sufficient to account for the observed phenomena, emphasizing the power of social contagion and memory distortion. Others, however, argue that the widespread and consistent nature of the effect necessitate a more radical re-evaluation of our understanding of reality and the possibility of non-linear temporal interactions.

The possibility of quantum phenomena influencing our perception of time and even our memories opens up new avenues of inquiry. Further research is needed to explore the potential connection between quantum mechanics, consciousness, and the Mandela Effect. This might involve investigating the correlation between specific quantum events and the reported instances of shared false memories, as well as exploring the role of collective consciousness in shaping our perception of reality. The implications of such research are far-reaching, challenging our fundamental assumptions about the nature of time, reality, and our place within the universe.

The potential for a deeper, more unified understanding of consciousness, time, and reality remains a compelling driver for this line of inquiry. The exploration of the Mandela Effect offers a unique window into the possible interplay between our macroscopic world and the strange, counter-intuitive realm of quantum mechanics. It compels us to question the very foundations of our understanding of time and reality, and to embrace a more open-minded approach to exploring the mysteries of the universe. The ongoing investigation of the Mandela Effect, therefore, is not merely a matter of understanding a peculiar cognitive phenomenon, but a potential key to unlocking a deeper comprehension of the cosmos and our place within its complex fabric. Whether we eventually find conclusive answers, the journey itself offers valuable insights into the boundaries of our current knowledge and the vast potential for future scientific exploration at the intersection of physics, psychology, and the enigmatic nature of time. The implications of such an endeavor may extend beyond our current understanding, potentially revolutionizing our conceptualization of reality and our very understanding of what it means to exist.

2 Parallel Universes and Multiverse Theories

The exploration of the Mandela Effect and its potential connection to non-linear time naturally leads us to consider the implications of multiverse theories. While the Mandela Effect itself remains a topic of ongoing debate and research, its perplexing nature compels us to explore theoretical frameworks that might provide plausible, albeit speculative, explanations. One such framework is the Many-Worlds Interpretation (MWI) of quantum mechanics, a theory that, while controversial, offers a particularly intriguing lens through which to view the possibility of parallel universes and their potential interaction with our own reality.

The MWI, initially proposed by Hugh Everett III in the 1950s, offers a radical departure from the traditional Copenhagen interpretation of quantum mechanics. The Copenhagen interpretation suggests that a quantum system exists in a superposition of states until a measurement is made, at which point the system "collapses" into a single definite state. The MWI, however, posits that the act of measurement does not cause a collapse of the wave function, but rather a branching or splitting of the universe. Each possible outcome of the measurement becomes a reality in its own separate universe, resulting in a continuously branching, ever-expanding multiverse.

In essence, the MWI suggests that every quantum event, from the decay of a radioactive atom to the interaction of subatomic particles, leads to the creation of new universes. These universes are not merely hypothetical constructs; they are considered to be equally real and physically existent, albeit inaccessible to us. This leads to a vast, almost incomprehensibly large multiverse containing countless parallel realities, each playing out a unique version of reality based on the myriad possible outcomes of every quantum event.

The implications of the MWI for our understanding of time are profound. In a linear conception of time, events unfold sequentially in a single, predetermined path. The MWI, however, challenges this view, suggesting that time itself might be a multi-branched, non-linear structure, with different universes experiencing different temporal pathways. This means that events that occur in one universe might not occur in another, or might occur in a significantly different order. This radical departure from linear temporality aligns with some of the more unconventional interpretations of the Mandela Effect, suggesting that perceived anomalies might be glimpses into alternate timelines or universes.

Consider the example of the misremembered "Luke, I am your father" dialogue from Star Wars. In the MWI framework, this could be interpreted as a fleeting interaction with a parallel universe where the dialogue differed slightly. A "bleed-through" from this parallel universe, possibly facilitated by some currently unknown quantum phenomenon or a consciousness-related process, might then manifest as a shared false memory among individuals who briefly experienced a sliver of that alternate reality. This isn't a complete shift into a new reality, but rather a transient intrusion of one universe upon another, leaving behind a trace in the form of altered memory.

Similarly, the misremembered logos and brand names associated with the Mandela Effect could be viewed as glimpses into universes where those branding elements are slightly different. Perhaps in one universe, the Fruit of the Loom logo did indeed feature a fruit bowl; in the MWI, this is not an impossibility, but rather just another branch in the multiverse. These shared misrememberings are not simply a matter of psychological error, but rather, a form of cross-universal interference, albeit subtle and fleeting.

The MWI, however, faces significant challenges. One major challenge is the lack of empirical evidence. While the theory is mathematically consistent with the framework of quantum mechanics, it remains extremely difficult, if not impossible, to directly observe or verify the existence of these parallel universes. The branching process described by the MWI occurs at the quantum level, a realm inaccessible to our current technology. We lack the tools to measure or detect the splitting of universes, let alone travel between them.

Moreover, the sheer scale of the multiverse proposed by the MWI is conceptually challenging. The number of parallel universes would be astronomical, potentially infinite, making it exceedingly difficult to develop testable predictions or devise experiments to verify the theory. It is important to note that MWI is a philosophical interpretation, not a theory in the usual sense that can be verified directly by experiment.

Despite these challenges, the MWI remains a compelling theoretical framework for understanding the potential relationship between quantum mechanics and the possibility of parallel realities. Its ability to potentially explain phenomena like the Mandela Effect, even speculatively, adds to its appeal. The idea that our perceived reality might be just one among countless others, each with its own unique timeline and set of events, stretches the boundaries of our understanding of reality.

The implications of the MWI extend far beyond the Mandela Effect. It raises fundamental questions about the nature of identity, free will, and the very essence of reality. If multiple versions of ourselves exist in parallel universes, each making different choices and experiencing different lives, what does it mean to be "me"? How can we reconcile the idea of free will with the deterministic nature of quantum mechanics, if every possible outcome becomes a reality in a separate universe?

Furthermore, the MWI challenges our understanding of causality. If events in one universe can indirectly influence events in another (as some interpretations of the Mandela Effect propose), how can we define causality in a multiverse context? The traditional understanding of cause and effect, so central to our scientific worldview, might need to be re-evaluated in light of the MWI's implications.

The exploration of the MWI and the multiverse, therefore, opens up a vast and unexplored landscape of philosophical and scientific possibilities. It is a challenging theory that pushes the boundaries of our conventional understanding of time, reality, and the universe. While the existence of parallel universes remains a topic of scientific and philosophical debate, the MWI provides a compelling theoretical framework for considering the possibility, and its potential implications for phenomena like the Mandela Effect are sufficiently intriguing to warrant further exploration. As our understanding of quantum mechanics deepens and our technological capabilities advance, we might eventually gain new insights into the nature of reality and the existence, or non-existence, of these parallel worlds. The continued study of these seemingly paradoxical concepts holds the key to unlocking a deeper, perhaps more unified, understanding of the cosmos and our place within its potentially infinite expanse. The quest to understand the mysteries of the universe is an ongoing journey, and the exploration of the

Many-Worlds Interpretation, however speculative, represents a significant step in that journey. It compels us to confront our ingrained assumptions about reality and to explore the uncharted territory of what lies beyond our current understanding. The potential rewards of this exploration are immense, potentially revolutionizing our understanding of the universe and our place within it.

The Many-Worlds Interpretation (MWI) of quantum mechanics, as explored in the previous section, provides a compelling, albeit speculative, framework for understanding the possibility of parallel universes. However, to fully appreciate the implications of MWI and its potential connection to phenomena like the Mandela Effect, we must delve into a crucial concept within quantum mechanics: quantum entanglement. This phenomenon, alongside the principle of non-locality, directly challenges our classical understanding of space and time, potentially bridging the gap between seemingly disparate universes.

Quantum entanglement describes a situation where two or more particles become linked in such a way that their fates are intertwined, regardless of the distance separating them. This linkage is not a matter of classical correlation; rather, it's a deeper, more fundamental connection that transcends spatial separation. Once entangled, the particles share a common fate, even if they are light-years apart. Measuring the properties of one entangled particle instantaneously influences the state of the other, regardless of the distance. This instantaneous correlation, defying the speed of light limitation imposed by Einstein's theory of special relativity, is what makes entanglement so profoundly strange and captivating.

The implications of entanglement for our understanding of non-locality are profound. Non-locality, in this context, refers to the idea that events can be correlated in ways that cannot be explained by local interactions – that is, interactions that are limited by the speed of light. In a classical Newtonian worldview, interactions are always localized; actions have a definite cause-and-effect chain that propagates through space at a speed no faster than light. Entanglement, however, shatters this classical view. The instantaneous correlation between entangled particles suggests that there's a deeper, non-local connection between them, a connection that transcends the limitations of space and time as we conventionally understand them.

This non-locality is not easily reconciled with our intuitive sense of reality. How can two particles, separated by vast distances, instantaneously influence each other? This phenomenon has been experimentally verified numerous times, leaving physicists with the perplexing task of trying to incorporate this apparent violation of locality into our existing framework of physics. Some interpret this non-locality as evidence for a deeper interconnectedness of reality, a cosmic interconnectedness that might explain the seemingly impossible connections proposed by some interpretations of the Mandela Effect. Could the shared false memories associated with the effect be the result of some form of subtle, quantum entanglement between the minds of multiple individuals, somehow accessing shared information across vast temporal or spatial scales? This is a highly speculative area, but it warrants consideration.

The concept of non-locality is further complicated by the uncertainty principle, another cornerstone of quantum mechanics. The uncertainty principle states that certain pairs of physical properties, like position and momentum, cannot be simultaneously known with perfect accuracy. The more precisely one property is known, the less precisely the other can be known. This inherent uncertainty in the quantum realm adds another layer of complexity to understanding entanglement and non-locality. The instantaneous correlation between entangled particles might be a consequence of this inherent uncertainty, suggesting that the particles are somehow connected in a way that transcends our classical understanding of cause and effect.

Numerous experiments have demonstrated the reality of quantum entanglement. One notable example is the Bell's theorem experiments. These experiments, designed to test the predictions of quantum mechanics against those of classical physics, repeatedly demonstrated that quantum mechanics accurately describes the behaviour of entangled particles, while classical physics fails. Bell's theorem demonstrated that no local realistic theory can reproduce the predictions of quantum mechanics for entangled particles. This reinforced the notion that quantum mechanics requires a non-local interpretation.

The connection between entanglement, non-locality and the multiverse theories, particularly the MWI, is a subtle yet potentially profound one.

If the MWI is correct, and every quantum event leads to the creation of a new universe, then entanglement could be interpreted as a form of communication or interaction between these parallel universes. Entangled particles, in this view, might not be simply correlated, but rather aspects of the same underlying reality, extending across multiple universes. The instantaneous correlation could be a reflection of a deeper connection, a connection that transcends the boundaries of individual universes.

This interpretation opens up fascinating possibilities. If entangled particles can interact instantaneously across vast distances within a single universe, could they also interact across different universes? Could this interaction explain the shared, seemingly anomalous memories associated with the Mandela Effect? It's crucial to acknowledge the highly speculative nature of these ideas. We are venturing far beyond the realm of currently verifiable science. However, the conceptual possibility is too compelling to ignore, prompting further investigation and theoretical exploration.

Furthermore, the role of consciousness in quantum entanglement remains an open question. Some physicists and philosophers suggest that consciousness might play a role in the collapse of the wave function, the moment when a quantum system transitions from a superposition of states into a single definite state. If consciousness is involved in this process, could it also play a role in entanglement and non-locality? Could conscious observation somehow facilitate the interaction between entangled particles, potentially bridging the gap between different universes or different points in time?

These questions bring us to the intersection of physics, philosophy, and even parapsychology. The potential link between consciousness, quantum entanglement, and the multiverse is a fertile ground for interdisciplinary research, a space where the boundaries of traditional scientific disciplines become increasingly blurred. It is important to approach these topics with a blend of rigorous skepticism and open-mindedness, acknowledging the limitations of our current understanding while maintaining a willingness to explore the implications of seemingly impossible phenomena.

The exploration of quantum entanglement and non-locality, within the framework of multiverse theories, represents a significant frontier in our quest to understand the nature of reality. While the answers remain elusive, the journey itself is immensely valuable. The very act of grappling with these seemingly paradoxical concepts forces us to challenge our ingrained assumptions and push the boundaries of our understanding of the cosmos and our place within it. The potential rewards of this exploration are substantial, potentially revolutionizing our view of the universe and ourselves. Whether or not we ultimately find definitive proof of parallel universes or inter-universal communication, the pursuit of these questions continues to illuminate the profound mysteries at the heart of existence. The journey of scientific inquiry, even into the most speculative realms, is a journey of discovery, a journey that enriches our understanding of the universe and our place within it.

String theory, at its core, attempts to reconcile the seemingly irreconcilable – quantum mechanics and general relativity. Quantum mechanics, incredibly successful in describing the subatomic world, paints a picture of reality governed by probabilities and uncertainties. General relativity, equally triumphant in its description of gravity and the large-scale structure of the universe, presents a universe governed by deterministic laws where spacetime is a smooth, continuous fabric. The problem is that these two theories, while incredibly powerful in their respective domains, are fundamentally incompatible at certain scales, particularly at the singularities found within black holes or at the very beginning of the universe.

String theory proposes a radical solution: Instead of viewing fundamental particles as point-like objects, it suggests that they are actually tiny, vibrating strings. These strings are not one-dimensional; they possess extra spatial dimensions beyond the three we experience – length, width, and height. These extra dimensions are thought to be compactified, curled up at incredibly small scales, invisible to our current observational capabilities. The different vibrational modes of these strings determine the properties of the particles we observe. A different vibrational pattern corresponds to a different particle: an electron, a quark, a photon, and so on.

The elegance of string theory lies in its potential to unify all fundamental forces of nature, including gravity, within a single, coherent framework. However, this elegance comes at a cost – the theory's mathematical complexity is daunting, and it has yet to yield testable predictions that could definitively confirm or refute it. Moreover, string theory leads to a breathtakingly expansive landscape of possible universes, often described as the "string theory landscape," with a vast number of potential solutions, each representing a different universe with different physical laws and constants. This theoretical landscape immediately introduces the concept of a multiverse – a reality composed of countless universes, each with its own unique characteristics.

This multitude of universes presents significant challenges to our understanding of time. In classical physics, time is considered a linear, absolute entity, flowing uniformly forward. However, string theory, with its numerous dimensions and universes, significantly complicates this picture. The extra spatial dimensions, often imagined as curled-up circles or other intricate geometrical shapes, interact with the temporal dimension in ways that are still not fully understood. One consequence is the potential for time to behave differently in different universes, possibly flowing at different rates or even reversing its direction in certain regions.

Consider the implications of a universe where time flows differently than in our own. Imagine a universe where time is cyclical, constantly repeating itself in a cosmic loop, or a universe where time progresses in discrete jumps rather than continuously. These possibilities, while deeply speculative, arise naturally from certain interpretations of string theory. Furthermore, the interactions between different universes within the multiverse could potentially lead to temporal anomalies or paradoxes. Could information, or even matter, somehow leak between universes, causing disturbances in the linear flow of time within our own universe?

The idea of time as a non-linear, multi-dimensional entity is further supported by certain interpretations of quantum mechanics, specifically the Many-Worlds Interpretation (MWI) which we discussed previously. Recall that in MWI, every quantum measurement causes the universe to split, creating multiple parallel universes, each representing a different possible outcome of that measurement. In this context, time could be seen as a branching, tree-like structure, with different branches representing different universes and different temporal pathways. This non-linear conception of time in MWI directly parallels the complex temporal possibilities suggested by string theory's multi-dimensional universe.

The implications of a non-linear time are profound, not just for physics but also for our understanding of consciousness and free will. If time isn't a single, unvarying flow, but a complex tapestry of multiple temporal streams, it challenges our conventional notions of causality and determinism. The possibility that events in one temporal stream might influence events in another opens up exciting avenues for philosophical speculation. Could our choices and actions in one temporal stream have unforeseen consequences in another? Could we influence the past or the future by interacting with other temporal streams?

These questions remain firmly in the realm of speculation, but their consideration forces us to question deeply held assumptions about the nature of reality. The notion of a multiverse with a complex, non-linear time suggests a reality far richer and more intricate than the classical worldview allows. String theory, while still under development and requiring further empirical validation, offers a compelling framework for exploring these possibilities. It presents a pathway to understanding the universe not merely as a collection of objects interacting in space and time, but as a dynamic, interconnected system with multiple temporal dimensions and an unimaginably vast number of parallel universes.

The challenge for physicists is not just to develop the mathematical framework of string theory further, but also to find ways to experimentally test its predictions. The scales involved are so incredibly small and the energies so incredibly high that direct experimental verification remains a formidable challenge. However, indirect evidence, such as subtle anomalies in cosmological observations or new theoretical insights, may gradually accumulate over time, offering tantalizing hints about the validity of string theory and the existence of a multiverse with a non-linear time.

Another significant challenge lies in developing intuitive and accessible ways to understand these complex concepts. The mathematics of string theory is highly abstract, and even the conceptual framework can be difficult for non-specialists to grasp. However, it is vital to bridge this gap between the esoteric realm of theoretical physics and the public's understanding of the universe. The implications of string theory and the multiverse are far-reaching and profoundly philosophical, affecting our perception of ourselves and our place in the cosmos. The journey towards a deeper understanding of reality is a collaborative one, requiring not just rigorous mathematical development but also creative explanations that make these awe-inspiring concepts accessible to a wider audience.

The search for evidence supporting string theory and the multiverse is an ongoing and challenging endeavor. But the potential rewards are enormous. The discovery of extra spatial dimensions and a non-linear time would revolutionize not only physics but also our understanding of the universe's fundamental nature, our perception of causality, and even the possibility of communication or interaction with parallel universes. The implications extend far beyond the scientific realm, reaching into philosophy, theology, and even our everyday perceptions of reality. It is a quest that stretches the boundaries of human imagination, constantly pushing us to redefine our understanding of what is possible and what is real. The intricate dance between string theory and the complexities of time suggests a universe of breathtaking scale and profound mystery, a universe whose exploration promises to reveal wonders beyond our current comprehension. The journey, as always in scientific inquiry, is as important as the destination, challenging us to expand our knowledge and reshape our perspectives on the nature of existence itself. The possibility of a non-linear time, woven into the fabric of a multi-dimensional reality, opens up horizons that were once considered the exclusive realm of science fiction, inviting us to embark on an intellectual adventure of unprecedented scope and depth.

The implications of a multiverse, as suggested by string theory, extend far beyond the mere existence of multiple universes. It deeply challenges our understanding of time itself, moving beyond the linear progression we experience daily. General relativity, while seemingly at odds with quantum mechanics at the fundamental level, provides a fascinating theoretical framework for understanding the potential for non-linearity: closed timelike curves (CTCs). These CTCs are essentially pathways through spacetime that loop back on themselves, creating the theoretical possibility of time travel and, crucially for our discussion, time loops.

Imagine a path through spacetime that, instead of extending indefinitely into the future, curves back upon itself, forming a closed loop. This is the essence of a CTC. Within such a loop, an event could both precede and succeed itself, creating a paradoxical situation that fundamentally challenges our intuitive understanding of causality. Consider the classic "grandfather paradox": if one were to travel back in time and prevent their own grandfather from meeting their grandmother, they would never have been born, thus negating the possibility of their journey into the past. This paradoxical scenario highlights the inherent tension between the linear conception of time and the possibility of CTCs.

The existence of CTCs, while purely hypothetical within our current understanding of the universe, is not ruled out by the equations of general relativity. Indeed, some solutions to Einstein's field equations predict the existence of such curves, particularly in extreme gravitational environments like those found near rotating black holes. These theoretical possibilities, however, raise a plethora of perplexing questions. How would the laws of physics behave within a CTC? Would the conservation of energy and momentum still hold? Could information be paradoxically created or destroyed within these loops?

One potential resolution to the paradoxes presented by CTCs is the concept of "self-consistency." This principle suggests that only those time loops that are self-consistent, i.e., those that do not lead to paradoxes, are physically permissible. If a time traveler were to attempt to alter the past in a way that would lead to a paradox, the universe, according to this principle, would somehow prevent that alteration from occurring. Perhaps the laws of physics themselves would subtly shift to ensure self-consistency, effectively rendering the paradoxical action impossible. This approach avoids the logical contradictions inherent in many time travel scenarios but leaves much unexplained concerning the mechanics of this self-consistency enforcement.

Another approach to resolving the paradoxes of CTCs involves the idea of parallel universes or branching timelines. Upon entering a CTC, the time traveler might not actually alter their own past but instead create a branching timeline, a new universe entirely, where the changes they make occur. This avoids contradictions within the original timeline, preserving the linearity of the past from the perspective of the original universe. This interpretation, which echoes the Many-Worlds Interpretation of quantum mechanics, provides a way to accommodate the potential for time travel without violating causality. Each time loop creates a new, self-consistent universe.

However, this multiverse approach presents its own challenges. It requires the acceptance of a vast, possibly infinite number of universes, each with its own unique history and timeline. While not fundamentally incompatible with string theory, it pushes the bounds of our ability to test and verify such a vast and complex cosmic architecture. How would we ever observe or interact with these alternative timelines born from time loops? The sheer scale of the multiverse, populated with countless branching universes and possible timelines, defies our current scientific capabilities to explore it in a direct manner.

The implications of CTCs and time loops extend beyond the purely physical realm. The very possibility of such phenomena deeply challenges our philosophical understanding of free will and determinism. If time is not a simple linear progression but can loop back on itself, it raises questions about the extent to which our actions are predetermined. Does the existence of CTCs imply a form of predetermination, where the future is somehow already fixed, even if we perceive it as open to change? Or does the existence of these loops suggest that our choices truly have agency, capable of shaping the flow of time itself within the confines of the self-consistency principle?

The exploration of CTCs and time loops also intersects with the study of consciousness. Some theories propose that consciousness itself might play a crucial role in resolving the paradoxes associated with CTCs. Perhaps consciousness itself is the underlying mechanism that enforces self-consistency within time loops. This notion ties into a broader discussion about the nature of reality, questioning whether time is an objective feature of the universe or a subjective construct of our consciousness. If time is perceived differently by different observers or even within different universes, it could further complicate our understanding of causality and paradoxes.

The study of CTCs, while deeply theoretical and often speculative, serves as a vital frontier in our understanding of spacetime and time itself. The challenges presented by these paradoxical possibilities push the boundaries of both physics and philosophy, leading to new avenues of inquiry into the nature of reality and our place within the cosmos. The interplay between general relativity, quantum mechanics, string theory, and the conceptual framework of time loops highlights the deeply intertwined nature of these seemingly disparate fields of study.

Moreover, the implications of time loops extend to the possibility of advanced civilizations possessing technologies capable of manipulating spacetime. The creation of CTCs might represent an ultimate technological achievement, offering the possibility of accessing vast amounts of information or resolving seemingly intractable problems by accessing different points in a temporal loop. Of course, such possibilities are firmly within the realm of science fiction, yet their exploration stimulates our imagination and challenges us to consider the implications of advanced civilizations operating on vastly different technological and philosophical levels compared to our own.

While we lack the empirical evidence to definitively confirm the existence of CTCs or time loops, exploring this concept allows us to confront the limitations of our current understanding of spacetime. It forces a reevaluation of the fundamental assumptions underpinning our models of the universe. This pursuit extends to the very nature of time itself, asking whether it is an absolute, universal parameter or a more malleable aspect of reality, potentially subject to manipulation or distortion under extreme conditions. The journey of scientific exploration, fueled by theoretical speculation and the desire for a more profound understanding of the cosmos, is an iterative process, pushing us ever closer to a more comprehensive and possibly unexpected image of the reality we inhabit.

The existence of a multiverse, with its attendant possibilities of parallel universes and branching timelines, coupled with the theoretical framework of closed timelike curves, opens up a panorama of possibilities regarding the nature of time and the structure of reality. While many of these possibilities remain deeply speculative and firmly within the realm of theoretical physics, their exploration provides a fertile ground for scientific advancement, philosophical inquiry, and an expansion of our understanding of the cosmos and our place within it. The inherent complexity of these topics challenges us to think creatively and develop new conceptual frameworks for accommodating concepts that defy our ingrained intuitions about time, causality, and the nature of reality itself. The pursuit of these seemingly paradoxical ideas ultimately serves to refine our understanding of the universe, pushing the boundaries of our scientific knowledge and philosophical imagination.

The theoretical frameworks of parallel universes and the multiverse, as discussed previously, have profoundly impacted popular culture, inspiring countless works of science fiction and fantasy. These fictional explorations often serve as thought experiments, pushing the boundaries of scientific plausibility and exploring the philosophical and ethical implications of a reality far more complex than our own. Analyzing these depictions provides valuable insight into how the public perceives and interprets these complex scientific concepts.

One of the earliest and most influential portrayals of parallel universes in fiction is found in the works of H.G. Wells. While not explicitly using the term "parallel universe," his concept of time travel, as depicted in "The Time Machine," implicitly suggests the existence of multiple timelines or realities. The idea that changes made in the past could create diverging realities, each representing a unique possible future, resonates strongly with the modern understanding of branching timelines and the Many-Worlds Interpretation of quantum mechanics. Wells' imaginative explorations laid the groundwork for numerous subsequent fictional interpretations of alternate realities.

The concept of parallel universes truly blossomed in the 20th and 21st centuries, fueled by the advancement of scientific understanding. Robert Heinlein's "The Number of the Beast," for instance, presents a complex and richly detailed multiverse where characters travel between different universes, encountering wildly different versions of reality, reflecting the infinite possibilities inherent in multiverse theories. Heinlein's intricate storytelling successfully integrates complex scientific concepts into a captivating narrative, demonstrating the creative potential of exploring the multiverse in fiction.

The Marvel Cinematic Universe (MCU) provides a particularly compelling example of parallel universes in popular culture, with its extensive exploration of the multiverse across its numerous films and television series. The MCU's depiction of the multiverse, while often simplified for the purpose of storytelling, effectively communicates the potential for vastly different versions of reality to exist simultaneously. The concept of alternate timelines, different versions of familiar characters, and the potential for interdimensional travel are explored extensively, engaging viewers with a visceral depiction of the multiverse's vastness and the potential for dramatic conflicts arising from interactions between these parallel realities.

The television series "Sliders" offers a more straightforward take on the multiverse, with characters traversing various parallel Earths, each with its own unique history and societal structures. The show cleverly uses the concept of parallel universes to explore "what if" scenarios, providing audiences with a glimpse into alternate historical paths and societal developments. The "Sliders" concept effectively illustrates the diversity inherent in the multiverse concept, emphasizing the vast range of potential realities that could exist.

Other notable examples include the "Back to the Future" trilogy, which, although primarily focused on time travel, implicitly incorporates elements of parallel universes through the notion of altered timelines. The changes made in the past create branching timelines that diverge from the original reality, similar to the branching timelines proposed in multiverse theory. These popular film series effectively showcase the potential for causality issues stemming from interactions with multiple timelines, reflecting the inherent complexities discussed in the scientific literature.

Video games have also embraced the concept of parallel universes, with games like "Baldur's Gate 3" employing a variety of branching narratives and alternate possibilities, allowing players to influence the events of the game and potentially create different timelines through their choices, thereby engaging them in the consequences of their actions in shaping their in-game reality.

The use of parallel universes in fiction often serves a dual purpose. It provides a compelling narrative device, allowing for creative storytelling and exploring "what if" scenarios, but it also functions as a tool for exploring profound philosophical questions about identity, free will, and the nature of reality itself. The exploration of alternate realities in fiction challenges viewers and readers to confront the implications of such concepts. Do our choices matter if an infinite number of alternative realities exist? Does the existence of a multiverse undermine the significance of our own individual experiences? These are questions that frequently arise when considering the implications of parallel universes.

Popular culture often simplifies or alters the scientific concepts for dramatic purposes. The accurate representation of scientific theories can sometimes be secondary to the creation of an engaging narrative. Yet, even these simplified depictions can stimulate critical thinking and public engagement with complex scientific topics. The numerous fictional explorations of parallel universes have undoubtedly broadened the public's awareness of these theoretical concepts, making them more accessible and engaging for a wider audience. The widespread popularity of these fictional representations highlights the intrinsic human fascination with the unknown and the potential for a reality far beyond our own.

The fictional depictions of parallel universes in popular culture are rarely a literal translation of specific scientific theories. They often borrow concepts and themes from various multiverse interpretations, blending them with creative liberties to create compelling narratives. This creative adaptation serves as a valuable tool for public engagement with scientific ideas, offering a less abstract and more emotionally resonant way to explore the implications of complex concepts such as string theory, the Many-Worlds Interpretation, and the theoretical possibilities of closed timelike curves. The success of these fictional works illustrates the power of storytelling in disseminating scientific knowledge and sparking public curiosity about the mysteries of the universe.

Ultimately, the relationship between parallel universes in popular culture and the scientific theories is symbiotic. The scientific theories provide the conceptual framework for the fictional explorations, while the fictional explorations, in turn, offer a broader audience access to these concepts, inspiring further scientific inquiry and stimulating public discourse on the nature of reality and the limits of human understanding. The ongoing dialogue between science and fiction concerning the multiverse is a testament to the enduring human fascination with the cosmos and our place within it. The continued exploration of these ideas, both in scientific research and popular culture, promises to deepen our understanding of the universe and its many possibilities. The journey of understanding continues, fueled by both rigorous scientific inquiry and the boundless creativity of the human imagination. The interplay between these two domains enriches our knowledge and challenges our assumptions about the nature of reality itself.

3 Time Travel and its Paradoxes

The exploration of parallel universes and the multiverse naturally leads us to the thorny thicket of time travel paradoxes. While the concept of traversing through time holds an irresistible allure, woven into the fabric of countless science fiction narratives, the very possibility raises profound questions about causality and the nature of reality itself. At the heart of these conundrums lies the infamous Grandfather Paradox, a thought experiment that vividly illustrates the potential inconsistencies inherent in backward time travel.

The Grandfather Paradox, in its simplest form, poses a seemingly insurmountable challenge. If one were to travel back in time and prevent their own grandfather from ever meeting their grandmother, the traveler's own existence would become impossible. This creates a causal loop, a self-contradictory scenario where the cause (the time traveler's actions) prevents the effect (the time traveler's birth). This fundamental contradiction has fueled decades of debate among physicists, philosophers, and science fiction enthusiasts alike.

The paradox isn't merely a whimsical thought experiment; it strikes at the core of our understanding of causality – the principle that every effect has a cause, and that causes precede effects in a linear fashion. The Grandfather Paradox seemingly violates this fundamental principle, implying that time travel to the past is either impossible or requires a fundamental revision of our understanding of causality.

One proposed resolution involves the concept of parallel universes or branching timelines, a notion gaining traction within multiverse theories. The argument posits that when a time traveler alters the past, they don't change their *own* timeline but instead create a new, branching timeline. In this interpretation, the time traveler's actions in the past create a new, separate reality distinct from their original timeline. In this new reality, the grandfather might never have met the grandmother, but the time traveler's original timeline remains intact, preserving their existence.

This interpretation elegantly sidesteps the paradox. The act of time travel doesn't erase the time traveler's existence but rather creates a divergence in reality, resulting in a multitude of potential futures. The traveler might experience a different past and present in this new timeline, but their original timeline remains unchanged. This echoes the Many-Worlds Interpretation of quantum mechanics, where every quantum measurement causes the universe to split into multiple branches, each representing a different possible outcome.

However, the multiverse solution, while seemingly elegant, introduces its own set of complexities. The sheer number of possible universes that could be created through time travel is staggering, potentially leading to an infinite regress of branching timelines. Managing and comprehending this infinite proliferation of realities presents a significant challenge to our intuitive understanding of the universe.

Beyond the Grandfather Paradox, other paradoxes arise within the framework of time travel. The Bootstrap Paradox, for example, presents a scenario where an object or piece of information exists without a clear origin. Imagine a time traveler transporting a book back in time, where it eventually inspires a writer who unknowingly copies it, leading to its existence in the present. In this scenario, the book's origin becomes circular, without a clear starting point, challenging our linear understanding of cause and effect.

Another perplexing scenario is the Predestination Paradox, where the time traveler's actions in the past were always destined to occur. In this instance, the time traveler's journey is not a matter of free will, but rather a predetermined event within the fabric of time itself. This interpretation suggests a deterministic view of reality, where free will is an illusion, and all events are preordained. This raises profound philosophical questions about human agency and the role of choice in our lives.

The Novikov self-consistency principle attempts to reconcile these paradoxes. This principle asserts that time travel is governed by laws of physics that only permit self-consistent timelines. In other words, the laws of physics, or some yet-unknown force, prevent paradoxes from occurring. Any attempt to create a paradox would be thwarted by some yet-unknown mechanism, ensuring the overall consistency of the spacetime continuum. While appealing in its simplicity, it lacks a concrete physical mechanism to explain this self-consistency.

The implications of these paradoxes extend far beyond the realm of theoretical physics. They challenge our fundamental understanding of free will. If time travel is possible, and if past events are mutable, does this imply that our choices and actions have no real impact? Or, conversely, if our actions are predetermined, does it negate the very notion of free will? These philosophical questions continue to stimulate discussion among ethicists and philosophers, pushing the boundaries of our understanding of human agency and responsibility.

Moreover, the concept of time travel introduces profound ethical considerations. If time travel were possible, could it be used to alter the course of history, potentially leading to unintended and disastrous consequences? Or, if it's merely a journey through multiple timelines, does the responsibility for altering one's own timeline outweigh the potential consequences of altering another? These questions raise complex ethical issues, suggesting the need for careful consideration before contemplating such endeavors.

The exploration of time travel paradoxes isn't simply an exercise in intellectual curiosity; it forces us to confront the limitations of our current understanding of physics and the universe. Our intuitive understanding of time, based on our everyday experience, may not be adequate to grapple with the complexities of time travel and the implications of altering the past. The search for resolutions to these paradoxes necessitates a deeper understanding of spacetime, causality, and the fundamental nature of reality.

The Grandfather Paradox, and the other paradoxes associated with time travel, reveal the inherent limitations of our current theoretical frameworks. While the multiverse theory offers a potential solution, it introduces further complexities, highlighting the vast unknowns that remain in our understanding of the cosmos. The ongoing exploration of these paradoxes pushes the boundaries of physics, philosophy, and our understanding of the universe and our place within it. The quest to unravel these mysteries drives scientific inquiry and inspires countless narratives in fiction, ensuring that the allure of time travel and the challenges it presents remain a topic of ongoing fascination and debate. The continued engagement with these paradoxes is vital, not only for a more complete understanding of the universe but also for a deeper understanding of our own place within its vast and complex tapestry of existence. The challenge posed by time travel paradoxes is not just a matter of solving equations but of confronting the philosophical implications of a reality that might be far stranger and more intricate than we currently comprehend. Only through continuous exploration can we hope to unravel the mysteries and complexities that time travel presents. The interplay between rigorous scientific investigation and the boundless creativity of human imagination remains the driving force behind our quest to decipher the enigmas of time and space.

The Bootstrap Paradox, unlike the Grandfather Paradox which focuses on the alteration of the past, presents a different kind of temporal conundrum: a scenario where an object or piece of information seemingly exists without a discernible origin. It's a circular causality problem, a closed loop where the cause and effect are intertwined in such a way that neither seems to precede the other. Imagine, for instance, a time traveler bringing back a copy of Shakespeare's Hamlet to Elizabethan England. This copy then somehow influences a young playwright, who, unbeknownst to himself, essentially transcribes it, creating the "original" Hamlet that was later transported back in time. Where, then, did Hamlet truly originate? The answer, within the paradox, is nowhere and everywhere at once. There is no clear point of origin. The book exists, yet its creation seems to lack a genuine, linear cause-and-effect chain.

This paradox challenges our deeply ingrained sense of linear causality, the bedrock upon which our understanding of the universe is largely built. We are accustomed to a world where effects follow causes in a clear, sequential manner. The Bootstrap Paradox throws this into disarray, suggesting a reality where objects can emerge from a self-referential loop, existing without a definitive beginning. This poses fundamental questions about the nature of creation and existence itself. Is it possible for something to exist without a clear causal antecedent? Does this imply a form of temporal self-creation, a universe where things simply "are" without needing a prior cause?

The implications of the Bootstrap Paradox extend beyond mere objects. Consider a piece of information, a scientific theory, or even a musical composition. Imagine a time traveler transporting a revolutionary mathematical theorem back to a period before its supposed "discovery." This theorem then influences mathematicians of the past, leading to its eventual rediscovery, or rather, its "original" formulation, in the future from which the traveler initially came. This would create a self-contained informational loop, raising questions about intellectual property, originality, and the very nature of intellectual creation. Did the original discoverer of the theorem truly discover it, or did they merely act as a conduit for information already existing in a self-contained temporal loop?

Some physicists suggest that the Novikov self-consistency principle, mentioned in relation to the Grandfather Paradox, might also offer a solution to the Bootstrap Paradox. This principle posits that the laws of physics somehow prevent paradoxes from occurring. In the context of the Bootstrap Paradox, this might mean that the existence of the object or information is inherently consistent with the overall spacetime structure. The laws of physics, or some as-yet-unknown force, might somehow ensure that the object or information has always existed, even if its origin seems circular. However, this interpretation still leaves many questions unanswered. What is the underlying mechanism that ensures self-consistency? How does this mechanism operate within a universe potentially governed by multiple timelines or interpretations of quantum mechanics?

The multiverse theory, which offers a potential escape route for the Grandfather Paradox, could also be applied to the Bootstrap Paradox. One could argue that the object or information exists in multiple universes simultaneously. The time traveler merely transports it from one universe to another, not creating it but rather transferring it between different branches of reality. In this scenario, the object or information might have different origins across different universes, effectively sidestepping the problem of circular causality. This theory, while appealing in its avoidance of contradictions, also introduces the complexity of navigating and understanding an infinite multitude of parallel realities.

Furthermore, the Bootstrap Paradox challenges our understanding of free will. If an object or piece of information can exist without a clear origin, does this suggest that our actions and creations are not entirely of our own making? Could our thoughts, ideas, and even our identities be preordained within a self-contained temporal loop? The implication here extends beyond the purely physical into the realm of the metaphysical, blurring the line between determinism and free will. This raises profound philosophical questions about the nature of consciousness and the extent to which we truly control our own destinies.

The possibility that information itself can exist in a self-referential temporal loop also raises intriguing questions for information theory. Is information truly created, or is it simply discovered, existing in some form of timeless informational reservoir? This challenges the conventional understanding of information as something that is created and destroyed. The Bootstrap Paradox suggests that information might be eternally present, circulating through time in a self-contained loop, perpetually influencing and being influenced by itself.

Beyond the strictly scientific, the Bootstrap Paradox has interesting implications for art and creativity. If a work of art, a piece of music, or a literary creation could emerge from a temporal loop, what does this say about originality and authorship? Does it diminish the role of the artist or creator, reducing them to mere conduits for pre-existing information? Or, conversely, does it elevate the artist, transforming them into participants in a timeless creative process? This raises significant questions about the relationship between the artist and their art, and the very nature of artistic inspiration.

One might consider the hypothetical case of a musician discovering a piece of music from the future, unconsciously incorporating it into their own compositions. This piece of music, later played to the future from whence it originally came, could then be said to have no actual origin point. This challenges our understanding of musical creation, blurring the lines between inspiration and imitation, and raising questions about the true nature of artistic originality.

The Bootstrap Paradox ultimately forces us to confront the limitations of our current understanding of time, causality, and the nature of reality itself. It challenges the very foundations of our scientific and philosophical frameworks, forcing us to re-evaluate our assumptions about the universe and our place within it. The solution, if a solution exists, may lie not in a simple scientific formula but in a fundamental shift in our understanding of how the universe works, a re-evaluation of the very concepts of time, causality, and creation. The exploration of this paradox, therefore, is not merely an intellectual exercise; it is a journey into the heart of some of the most profound mysteries of existence.

The paradox's enduring appeal stems from its ability to challenge our deepest-held assumptions about the nature of reality. It pushes us beyond the comfortable confines of linear causality and forces us to consider the possibility of a universe far more complex and unpredictable than we have hitherto imagined. By confronting the paradoxes inherent in time travel, we are not simply engaged in a purely intellectual exercise; we are actively engaged in the expansion of human understanding, a quest to unravel the deeper mysteries of the universe and our place within its intricate tapestry. The Bootstrap Paradox, in its seemingly simple formulation, provides a compelling gateway to deeper exploration of some of the most fundamental questions regarding existence, consciousness, and the very fabric of reality itself. Its resolution, or even a closer understanding of its implications, promises to bring us closer to a more comprehensive and nuanced perspective on the nature of time and the universe that we inhabit. The exploration will continue, pushing the boundaries of science and philosophy, and enriching our comprehension of the cosmos and our own intriguing place within it.

Einstein's theories of relativity, particularly his special and general theories, form the cornerstone of most theoretical discussions on time travel. Special relativity, published in 1905, introduced the concept of spacetime – a four-dimensional framework where three spatial dimensions are interwoven with time. A crucial consequence is time dilation: the faster an object moves relative to an observer, the slower time passes for that object from the observer's perspective. This isn't a subjective experience; it's a predicted and experimentally verified phenomenon. High-speed particles in particle accelerators, for example, experience time dilation, their internal clocks ticking slower than those of stationary observers.

While special relativity hints at the possibility of time travel through extreme velocities, it doesn't offer a practical mechanism. General relativity, published in 1915, provides a more nuanced perspective. This theory describes gravity as the curvature of spacetime caused by mass and energy. Massive objects warp spacetime, altering the flow of time. The closer an object is to a massive body, the slower time passes for it relative to an observer further away. This effect, known as gravitational time dilation, has been experimentally verified using highly precise atomic clocks at different altitudes.

General relativity opens up intriguing possibilities for time travel. One theoretical concept is the wormhole, a hypothetical tunnel through spacetime connecting two distant points. These hypothetical structures, predicted by Einstein's field equations, are essentially shortcuts through the fabric of spacetime. Traveling through a wormhole could, theoretically, allow for journeys across vast distances or even into the past or future. However, the existence of wormholes remains purely theoretical. Their stability poses a significant challenge. The immense gravitational forces involved would likely collapse any naturally occurring wormhole, rendering it impassable.

Furthermore, the creation of an artificial wormhole would require manipulating spacetime on a scale far beyond our current technological capabilities. Negative mass-energy, a substance with properties opposite to those of ordinary matter, is often postulated as a necessary component for stabilizing a wormhole. However, the existence of negative mass-energy is still hypothetical. Even if it did exist, harnessing and manipulating it would present insurmountable technological hurdles.

Another theoretical approach to time travel, albeit equally challenging, involves the manipulation of cosmic strings. These are hypothetical one-dimensional topological defects in spacetime, predicted by some cosmological models. Cosmic strings possess immense density and gravitational fields. Certain configurations of cosmic strings, according to some theoretical calculations, could create closed timelike curves (CTCs). A CTC is a path through spacetime that loops back on itself, allowing an object to return to its own past. Theoretically, traversing a CTC could enable time travel. However, the existence of cosmic strings is still unconfirmed, and manipulating them to create CTCs would present immense technological challenges, even beyond those associated with wormholes.

Even if we were to overcome the technological hurdles of creating a stable wormhole or manipulating cosmic strings, there remain profound conceptual obstacles. The paradoxes discussed earlier – the Grandfather Paradox and the Bootstrap Paradox – are not merely philosophical thought experiments. They highlight inherent inconsistencies that could arise if time travel were possible. The potential for altering the past and disrupting the causal chain of events poses profound questions about the nature of free will, determinism, and the very fabric of reality.

The Novikov self-consistency principle, while providing a potential solution to some paradoxes by suggesting that the laws of physics somehow prevent paradoxes from occurring, doesn't fully address the underlying conceptual issues. It merely suggests a mechanism that maintains consistency, but doesn't explain how that consistency is ensured or what its implications are for our understanding of causality and free will. Similarly, the multiverse theory, while providing an alternative framework in which paradoxes might be avoided, introduces its own set of complexities, requiring the consideration of an infinite number of parallel universes.

The challenges of time travel extend beyond theoretical physics. The energy requirements for accelerating an object to near-light speed, necessary for significant time dilation, are astronomical. Even with futuristic energy sources, the engineering challenges of building a spacecraft capable of withstanding the immense forces involved would be staggering.

Beyond the purely physical aspects, the ethical implications of time travel are equally profound. The potential for altering the past, creating alternate timelines, or interfering with the course of history raises serious ethical considerations. The unintended consequences of such actions could be catastrophic. The responsibility associated with the power to manipulate time itself presents a significant moral dilemma.

In conclusion, while Einstein's theories of relativity offer tantalizing glimpses of the possibility of time travel, the practical and conceptual obstacles remain immense. The technological challenges associated with creating stable wormholes or manipulating cosmic strings are currently insurmountable. The paradoxes inherent in time travel highlight fundamental inconsistencies that would need to be resolved before time travel could be considered a feasible reality. Even if the technological and theoretical hurdles were overcome, the ethical implications would demand careful consideration. The pursuit of time travel, therefore, is not simply a matter of engineering but also a profound philosophical and ethical challenge. While the possibility remains a captivating subject of scientific inquiry and speculative fiction, the road to practical time travel remains long and potentially fraught with unforeseen difficulties. The journey into the depths of temporal physics, therefore, is likely to remain a fascinating and challenging pursuit for many years to come. The exploration of such theoretical concepts continues to push the boundaries of our understanding of the universe and our place within it. The quest to unravel the mysteries of time travel remains an engaging endeavor, enriching our understanding of physics, cosmology and the very nature of reality itself. The ongoing debate on the possibility and consequences of time travel will continue to stimulate scientific thought and philosophical inquiry, offering insights into the fundamental aspects of existence. The study of time travel, therefore, is not merely a pursuit of scientific curiosity but a profound examination of the universe and our own intriguing existence within it.

The inherent paradoxes associated with time travel, such as the Grandfather Paradox and the Bootstrap Paradox, have led to considerable debate among physicists and philosophers. These paradoxes highlight the potential for inconsistencies if time travel were possible, raising serious questions about causality and the very fabric of reality. One prominent attempt to address these inconsistencies is Stephen Hawking's Chronology Protection Conjecture.

Hawking, a towering figure in theoretical physics, proposed this conjecture not as a rigorously proven theory, but rather as a hypothesis suggesting that the universe might possess an inherent mechanism to prevent the creation of closed timelike curves (CTCs), the pathways through spacetime that would theoretically allow for time travel into the past. He argued that the laws of physics themselves might conspire to prevent paradoxes from occurring. This doesn't mean time travel is impossible, but rather that the universe might actively prevent scenarios that would lead to logical contradictions. The conjecture posits that the quantum effects near a CTC would become so strong that they would effectively destroy the CTC before it could form, preventing any paradoxes from arising. It's a universe operating under a principle of self-consistency.

Imagine attempting to build a time machine that could send you back to alter a significant past event. According to the Chronology Protection Conjecture, as you approached the point where your actions could create a paradox, the quantum fluctuations in spacetime would become increasingly intense. These fluctuations could manifest as increasingly powerful energy fields, intense gravitational forces, or even the spontaneous creation of particles and antiparticles, ultimately disrupting the structure of spacetime and preventing the completion of the time travel journey. The universe, in essence, would self-correct, preventing you from ever reaching the point where you could create a paradox.

This is not a claim that time travel is absolutely impossible, merely that the laws of physics might naturally prevent paradoxes. Hawking's suggestion hints at a universe that maintains its internal consistency, a sort of self-regulating system operating on a scale far beyond our current comprehension. The exact mechanisms by which this self-regulation might occur remain a subject of intense debate and speculation. Some theories propose that quantum gravity, a theoretical framework combining general relativity and quantum mechanics, plays a crucial role. The extreme conditions near a CTC could trigger quantum effects strong enough to prevent its formation.

The implications of the Chronology Protection Conjecture extend far beyond the simple avoidance of paradoxes. It speaks to the deep interconnectedness of spacetime and the fundamental laws that govern it. It suggests a universe where causality, while perhaps not entirely linear, is ultimately preserved. While seemingly paradoxical events might appear possible theoretically, the universe might prevent their actual occurrence through the operation of fundamental physical laws yet to be fully understood. This would suggest a degree of inherent stability in the fabric of spacetime, suggesting that the universe is far more resilient to drastic changes than some interpretations of time travel might suggest.

The concept is not without its critics. Some argue that the conjecture is too speculative, lacking a concrete physical mechanism to explain how this "chronology protection" is achieved. Others argue that it places arbitrary constraints on the laws of physics, effectively shutting down the possibility of time travel without offering a compelling alternative explanation for the apparent lack of time travelers in our present. The debate is further complicated by the lack of experimental evidence. Since we lack the ability to create or even detect CTCs, the conjecture remains largely untestable with our current technology. Moreover, the conjecture doesn't address all potential paradoxes. While it might address the Grandfather Paradox, other paradoxes, like the Bootstrap Paradox, where an object or information is created in a time loop, might still theoretically occur. However, it is worth noting that the absence of observable evidence does not necessarily constitute evidence of absence.

Considering the Chronology Protection Conjecture alongside other theories of time travel, such as those involving wormholes and cosmic strings, paints a richer, more nuanced picture. Wormholes, theoretical tunnels through spacetime, are often viewed as potential pathways for time travel. However, even if wormholes were to exist, their stability would be a major concern. The immense gravitational forces involved could easily collapse the wormhole, preventing passage. Similarly, cosmic strings, hypothetical one-dimensional topological defects in spacetime, are sometimes considered in theoretical time travel scenarios. Certain configurations of these strings might create CTCs, but their existence is purely hypothetical, and manipulating them to create stable time travel pathways would present an almost unimaginable technological challenge.

Comparing these theoretical avenues with the Chronology Protection Conjecture illustrates the inherent difficulties in bridging the gap between theoretical physics and practical reality. While the mathematics of general relativity may allow for the conceptual existence of CTCs, the conjecture suggests a self-correcting mechanism that might render them impossible. In this framework, the universe's fundamental laws might be more robust than we previously imagined, effectively acting as a self-imposed constraint on the potential for paradoxes. This could imply a deeper, more intricate level of organization within spacetime than we currently understand, reflecting a fundamental principle of self-consistency woven into the fabric of reality itself.

Moreover, the Chronology Protection Conjecture is deeply intertwined with the broader philosophical debate surrounding free will and determinism. If the universe actively prevents paradoxes, does this imply a degree of determinism? If our actions are constrained by the laws of physics in such a way that we cannot alter the past, does this limit our freedom of choice? These are profound questions that extend beyond the realm of theoretical physics and delve into the fundamental nature of existence itself.

The conjecture, therefore, acts not simply as a scientific hypothesis, but also as a catalyst for philosophical inquiry, raising fundamental questions about the relationship between physical laws and human agency.

The exploration of the Chronology Protection Conjecture continues to be a fertile ground for research and speculation. While it remains untested and potentially unprovable within our current framework of scientific understanding, it forces us to reconsider our fundamental assumptions about the nature of time, causality, and the universe itself. It is a testament to the power of theoretical physics to push the boundaries of our understanding, challenging us to confront the seemingly paradoxical nature of reality and to explore the intricate relationships between time, space, and the fundamental laws that govern them. The ongoing debate around its validity and implications serves to illuminate the depth and complexity of the universe, reminding us that the quest for knowledge is a continuous journey, filled with challenges, uncertainties, and unexpected discoveries. Ultimately, the conjecture's impact extends beyond the realm of theoretical physics, prompting broader discussions concerning the nature of reality, the limitations of human understanding, and the profound mysteries that continue to enshroud the universe. The seemingly insurmountable challenges of time travel serve not only as a testament to the limitations of our current technological capabilities, but also to the enduring power of scientific inquiry to push the boundaries of our understanding and to explore the fascinating paradoxes inherent in our universe's very structure.

The exploration of time travel, beyond the realm of theoretical physics and philosophical debate, has found a rich and fertile ground in popular culture. Science fiction, in particular, has extensively engaged with the concept, offering myriad depictions that reflect not only our scientific understanding (or lack thereof) but also our deepest anxieties and hopes regarding the malleability of time. These fictional narratives often serve as thought experiments, exploring the potential consequences of altering the past or influencing the future in ways that real-world science has yet to grapple with.

One of the most enduring and frequently explored themes is the Grandfather Paradox. Numerous science fiction stories have wrestled with the implications of traveling back in time and preventing one's own birth, creating a causal loop that defies logic. From the classic tale of a time traveler who accidentally kills his grandfather before his own father is conceived, to more nuanced variations involving alterations to seemingly minor events with unforeseen, ripple effects across time, the Grandfather Paradox serves as a cornerstone of many time travel narratives. The exploration of this paradox is often not merely a technical exercise but a reflection on the complexities of identity, free will, and the inherent linearity (or lack thereof) assumed within our temporal experience. The potential for the unraveling of one's own existence serves as a potent dramatic device, raising profound questions about the stability and consistency of our personal realities.

Beyond the Grandfather Paradox, the Bootstrap Paradox presents a different set of challenges. In this scenario, an object or piece of information exists without a clear origin. It is sent back in time, only to reappear as the very source of its own existence, creating a closed causal loop. This often manifests in stories involving an invention or technology that mysteriously appears in the past, seemingly springing into existence without a discernible inventor or precursor. This paradox highlights the potential for non-linear causality, suggesting that time may not be as unidirectional and linear as we commonly perceive it. The Bootstrap Paradox also touches upon questions of originality and the nature of creation itself, suggesting that some events or artifacts might exist independent of a singular point of origin. This concept stretches our understanding of causality and its relationship to temporal sequencing.

Popular culture often utilizes the concept of time travel as a narrative tool to explore historical events and their consequences. Many science fiction stories use time travel as a means to revisit pivotal moments in history, offering alternate perspectives and hypothetical scenarios. This can range from attempts to prevent historical catastrophes, such as the assassination of a pivotal political figure, to the exploration of what might have been had certain historical events unfolded differently. These narratives often serve as a reflection on our own understanding of history and its complexities. The potential to rewrite the past, however seductive, often reveals unintended and unforeseen consequences, highlighting the interconnectedness of historical events and the butterfly effect. The fictional exploration of these scenarios allows us to ponder "what ifs" and consider the delicate balance of causality on a grand scale.

The influence of temporal mechanics also extends to the character development in various science fiction works. Characters often struggle with the emotional and psychological implications of time travel. Experiencing the past or future can fundamentally alter their perception of self, their relationships with others, and their understanding of time itself. The emotional toll of witnessing events beyond their personal temporal existence, and the potential for altering those events, can lead to profound existential crises. These narratives often explore the psychological consequences of temporal displacement, the sense of dislocation and the challenges associated with maintaining one's identity across different time periods. The emotional weight of witnessing historical tragedies, or the frustration of attempting to influence seemingly immutable events, is often explored with compelling nuance.

Furthermore, the very act of building and operating a time machine is often a central narrative device in science fiction. The technological challenges, the ethical considerations, and the potential for misuse are all common themes. The process of constructing a time machine often involves overcoming seemingly insurmountable technological hurdles, forcing characters to push the boundaries of scientific knowledge and technological capabilities. These narratives often serve as metaphors for humanity's persistent pursuit of knowledge and its potential consequences. The potential for misuse is another frequent subject of exploration, highlighting the ethical considerations and dangers associated with manipulating the temporal flow. The responsibility of controlling such a powerful technology, and the potential for catastrophic consequences from misuse, are regularly examined within these fictional contexts.

The depiction of time travel in popular culture is not always confined to a strict adherence to scientific plausibility. Some science fiction works embrace fantastical and imaginative interpretations, defying conventional understandings of physics and causality. This allows for creative and imaginative explorations of temporal phenomena, exceeding the limitations imposed by current scientific knowledge. These narratives often serve as explorations of alternative realities, bending or breaking the rules of time to enhance narrative depth and to explore the limits of our imaginations. This flexibility in depicting time travel allows for a broader range of narratives, extending beyond the purely scientific and delving into the realm of imagination.

The diversity of representations across different media – novels, films, television series, and video games – underscores the enduring fascination with time travel. Each medium utilizes different narrative techniques and stylistic approaches, resulting in a diverse collection of interpretations that reflect the evolving understanding of time travel and its paradoxes. The medium influences the type of story that can be told, and the way in which time travel is presented can vary greatly depending on the platform. For example, the pacing of a narrative in a novel allows for a deeper exploration of psychological consequences than might be possible in a fast-paced action film.

In conclusion, the exploration of time travel in popular culture is far more than just a reflection of scientific progress. It is a vibrant tapestry woven with threads of philosophical contemplation, psychological exploration, and imaginative storytelling. These fictional representations provide a platform for grappling with the profound questions and anxieties that the concept of time travel inevitably evokes. From the seemingly simple paradoxes to the intricate entanglement of causality and free will, popular culture's engagement with time travel offers a valuable window into our understanding of time, our place within the universe, and our enduring fascination with the possibilities—and perils—of altering the past, present, or future. The continued exploration of time travel in fiction serves as a testament to our enduring curiosity and our relentless pursuit of understanding the fundamental nature of reality itself. The narratives themselves evolve alongside our own understanding of science, reflecting not just what we know, but also what we fear and what we hope for in the face of the unknown.

4 Ancient Astronaut Theories and Non-Linear Time

The preceding discussion of time travel, particularly its fictional portrayals, naturally leads us to consider its purported manifestations in the historical record. Ancient astronaut theories, a significant branch of fringe science, posit that extraterrestrial visitors have influenced or even directly shaped human civilization throughout history. These theories often incorporate concepts of advanced technology and potentially non-linear temporal mechanics, directly challenging conventional archaeological and historical interpretations. While largely dismissed by mainstream academia, the persistence and popularity of ancient astronaut theories offer a compelling case study in the human quest to understand our origins and place in the cosmos.

At the heart of these theories lies the argument that certain ancient structures and artifacts are too sophisticated to have been constructed with the technology available to the civilizations credited with their creation. Proponents often point to megalithic sites like the pyramids of Giza, the Nazca Lines of Peru, or Stonehenge in England as prime examples of inexplicable engineering prowess. The precision of these structures, the complex mathematical principles seemingly embedded in their designs, and the sheer scale of the undertaking lead some to postulate the involvement of beings with far more advanced technology than anything known to have existed at the time. These arguments frequently invoke the notion of advanced extraterrestrial civilizations possessing technologies beyond our current comprehension – technologies potentially including sophisticated energy sources, advanced construction techniques, and perhaps even forms of temporal manipulation.

The claim of extraterrestrial influence is frequently supported by interpretations of ancient texts and mythology. Proponents argue that many ancient myths and legends contain coded messages or symbolic representations of encounters with extraterrestrial visitors. They cite stories of gods descending from the heavens, flying chariots, and beings with superhuman abilities as potential accounts of ancient astronaut encounters.

Such interpretations, however, are often highly subjective and rely on creative interpretations of ambiguous textual evidence. The lack of concrete, universally accepted evidence remains a critical point of contention between proponents and skeptics. The subjective nature of such interpretations, vulnerable to confirmation bias and selective evidence gathering, undermines their scientific validity in the eyes of mainstream scholars.

Furthermore, ancient astronaut theories frequently engage with the concept of non-linear time, suggesting that our understanding of historical sequencing is incomplete. The proponents argue that advanced extraterrestrial civilizations may have interacted with human civilizations across vast stretches of time, perhaps even traveling back and forth through time, leaving behind traces of their influence that span millennia. This perspective challenges the linear progression of history typically assumed by conventional historical narratives, opening the possibility of discontinuous or even cyclical interactions between humanity and extraterrestrial intelligences. This perspective directly relates to the discussions surrounding time travel in the preceding chapter; instead of theoretical physics providing the framework, it's historical interpretation, albeit a highly contested one.

However, it's crucial to acknowledge the methodological challenges inherent in ancient astronaut theories. The absence of concrete physical evidence, such as unambiguous extraterrestrial artifacts or direct technological remains, represents a major weakness. Many of the arguments rely heavily on speculation, subjective interpretations, and a tendency to selectively emphasize data that supports the theory while disregarding contradictory evidence. Furthermore, the theories often lack rigorous scientific methodologies, relying instead on anecdotal accounts, unsubstantiated claims, and reinterpretations of existing historical and archaeological data.

A key criticism leveled against ancient astronaut theories is the lack of falsifiable predictions. A robust scientific theory should make predictions that can be tested and potentially proven false. Ancient astronaut theories, however, often lack this crucial element. The explanations provided frequently adapt to accommodate new evidence, making it difficult to disprove the theory regardless of any new discoveries. This adaptability, while seemingly flexible, ultimately hinders the theory's progress toward scientific acceptance. The very ambiguity inherent in the claims makes verification – and hence falsification – exceptionally difficult.

The cultural impact of ancient astronaut theories, regardless of their scientific validity, is undeniable. These theories have fueled popular culture, inspiring countless books, documentaries, and fictional works, captivating audiences with their narratives of ancient encounters and advanced extraterrestrial technologies. They cater to a deep-seated human desire to understand our origins, our place in the universe, and the potential existence of life beyond Earth. This cultural appeal speaks to a broader fascination with the unknown, a desire to find meaning and answers in narratives that challenge the established order.

However, it is essential to maintain a critical perspective. While the curiosity and desire to explore these questions are commendable, it's crucial to distinguish between speculative narratives and scientifically verifiable claims. The reliance on unsubstantiated claims, subjective interpretations, and the lack of rigorous scientific methodology detract from the credibility of these theories within the scientific community. The temptation to accept readily available, seemingly easy explanations should always be countered with rigorous critical thinking. The burden of proof lies with those making extraordinary claims, and the absence of concrete evidence necessitates a healthy skepticism.

The exploration of ancient astronaut theories highlights the inherent tension between imaginative storytelling and rigorous scientific inquiry. The compelling narratives they offer tap into fundamental human questions about our past, our place in the universe, and the potential for extraordinary encounters. Nevertheless, the lack of empirical evidence and the questionable methodology employed necessitate a cautious approach. The fascination with these theories shouldn't overshadow the importance of adhering to sound scientific principles, emphasizing verifiable evidence, and subjecting all claims to rigorous scrutiny. Understanding our past requires both imagination and a commitment to critical, evidence-based reasoning. The pursuit of knowledge should always prioritize rigorous investigation over captivating speculation, even when the latter presents itself in a compelling and imaginative manner.

The enduring appeal of ancient astronaut theories speaks to a broader human need for narratives that offer explanations for the mysteries of the past. While these theories often provide simple and exciting answers, they frequently lack the rigorous scrutiny and evidentiary support required for scientific acceptance. Therefore, it is vital to maintain a balance between imaginative exploration and critical evaluation. The quest to understand our origins and our place in the cosmos should be guided by both intellectual curiosity and a commitment to rigorous scientific methodologies. A healthy dose of skepticism, coupled with a thirst for knowledge, forms the bedrock of responsible scientific inquiry.

Furthermore, the exploration of ancient astronaut theories offers a valuable opportunity to reflect on the nature of evidence, interpretation, and the inherent limitations of our understanding of the past. The subjective interpretations of ancient texts and artifacts highlight the crucial role of critical analysis and the potential for bias in shaping our understanding of history. It also underscores the need for transparency and rigorous methodologies in scientific inquiry. By critically analyzing both the claims and the methodologies employed by proponents of these theories, we gain a deeper appreciation for the importance of robust scientific principles and the complexities of historical reconstruction. The discussion extends beyond the specific claims of ancient astronauts, serving as a broader reflection on the challenges and pitfalls of interpreting historical and archaeological data.

In conclusion, while ancient astronaut theories offer intriguing narratives and engage with profound questions about our origins, they remain, at present, outside the realm of established scientific consensus. The absence of concrete evidence, the reliance on subjective interpretations, and the lack of falsifiable predictions raise serious concerns about their scientific validity. However, the theories' enduring popularity reflects a fundamental human desire to understand our place in the cosmos and to seek answers to the enduring mysteries of our past. The critical analysis of these theories provides a valuable opportunity to reflect on the nature of evidence, interpretation, and the importance of rigorous scientific methodologies in understanding both the past and the possibilities of the future. The ongoing dialogue surrounding ancient astronaut theories underscores the dynamic interplay between imagination, speculation, and scientific rigor in our pursuit of knowledge.

Evaluating the evidence supporting ancient astronaut theories requires a nuanced approach, acknowledging both the provocative claims and the inherent limitations of the evidence presented. Proponents often point to architectural marvels like the pyramids of Giza, the precision of megalithic structures like Stonehenge, and the enigmatic Nazca Lines as evidence of technology far exceeding the capabilities of the civilizations conventionally attributed to their construction. The sheer scale of these projects, the sophisticated mathematical principles underlying their design, and the precision of their construction are indeed remarkable. However, simply asserting that these feats are "too advanced" for their time is insufficient to establish extraterrestrial intervention. This requires a detailed examination of the available evidence, considering alternative explanations, and applying principles of critical analysis.

One crucial tool for evaluating competing hypotheses is Occam's Razor – the principle that, given multiple explanations for a phenomenon, the simplest explanation with the fewest assumptions is generally preferred. While ancient astronaut theories offer a seemingly straightforward explanation – advanced extraterrestrial technology – they introduce a significant number of unproven assumptions. These include the existence of advanced extraterrestrial civilizations capable of interstellar travel, their willingness and ability to intervene in human history, and the specific mechanisms by which they achieved their purported feats of engineering. Each of these assumptions demands considerable evidence, none of which has been conclusively provided.

In contrast, alternative explanations, while potentially more complex, often rely on less speculative assumptions. For instance, the construction of the pyramids, once considered an insurmountable mystery, is now better understood through advancements in archaeology and engineering. The discovery of sophisticated tools, organizational structures, and a deep understanding of mathematics and physics within ancient Egyptian civilization casts significant doubt on the necessity of extraterrestrial involvement. Similarly, the construction techniques used at Stonehenge, while still partially enigmatic, are increasingly understood through ongoing archaeological research, suggesting incremental technological advancements and a complex social organization within Neolithic societies.

The Nazca Lines, often cited as evidence of extraterrestrial involvement due to their immense scale and precise geometric patterns, also offer alternative interpretations. Recent research suggests that the lines may have served ritualistic purposes, perhaps linked to astronomical observations or water management. While the exact function remains a matter of debate, the increasing evidence of sophisticated astronomical knowledge within pre-Columbian cultures provides a plausible explanation without invoking extraterrestrial intervention.

Moreover, the interpretations of ancient texts and mythology used to bolster ancient astronaut theories are frequently subjective and open to multiple interpretations. The symbolic language and metaphorical imagery present in ancient texts are often interpreted selectively to fit the narrative of extraterrestrial contact, neglecting other equally plausible interpretations. This selective approach to evidence is a significant weakness in the argumentation. A robust scientific approach requires a comprehensive and unbiased examination of all available evidence, rather than cherry-picking data to support a predetermined conclusion. The absence of clear, unambiguous evidence of extraterrestrial artifacts or technology within these texts remains a crucial flaw in the theory.

The concept of non-linear time, often invoked in ancient astronaut theories, presents further methodological challenges. The suggestion that extraterrestrial civilizations have interacted with humanity across vast stretches of time, potentially altering historical events, necessitates a radical re-evaluation of our understanding of chronology and causality. However, the lack of empirical evidence supporting non-linear temporal effects, combined with the theoretical challenges posed by such concepts within established physics, makes this a highly speculative claim. Extraordinary claims require extraordinary evidence, and the burden of proof rests with those proposing such a radical departure from our current understanding of time and history.

The proponents of ancient astronaut theories often point to the unexplained anomalies within the archaeological record as evidence for their hypotheses. However, the existence of unexplained phenomena does not automatically imply extraterrestrial intervention. The history of science is replete with examples of phenomena initially considered inexplicable that were later explained through advancements in knowledge and improved understanding of natural processes. The cautious approach of science dictates that we should first exhaust all plausible terrestrial explanations before resorting to extraterrestrial hypotheses. Unidentified or unexplained anomalies should be considered as open questions requiring further investigation, not automatically attributed to the intervention of advanced extraterrestrial civilizations.

A further criticism relates to the inherent difficulty in falsifying ancient astronaut theories. A good scientific theory should be falsifiable; it should make specific predictions that, if proven false, would invalidate the theory. Ancient astronaut theories, however, often adapt to accommodate new evidence, rendering them virtually unfalsifiable. This lack of falsifiability is a significant impediment to their scientific acceptance. A theory that can explain anything effectively explains nothing. True scientific progress depends on the ability to test and potentially refute hypotheses, leading to a refined and more accurate understanding of the world.

In conclusion, while the allure of ancient astronaut theories lies in their ability to provide seemingly simple explanations for complex phenomena, their lack of robust evidence, reliance on subjective interpretations, and inherent unfalsifiability render them inadequate from a scientific perspective. The application of Occam's Razor, the exploration of alternative explanations, and the rigorous application of scientific methodology are crucial for navigating the complexities of the past and discerning between imaginative narratives and verifiable historical accounts. While the questions of our origins and potential extraterrestrial contact remain compelling, it is paramount to adhere to the principles of sound scientific inquiry, ensuring that extraordinary claims are backed by extraordinary evidence. A balanced approach, embracing curiosity while maintaining a healthy dose of skepticism, is essential in the pursuit of knowledge. The mysteries of the past should inspire further research, employing robust methodologies and avoiding premature conclusions based on speculation alone.

The preceding discussion highlighted the limitations of ancient astronaut theories when viewed through the lens of conventional scientific methodology. However, the inherent appeal of these theories lies not solely in the possibility of extraterrestrial visitation, but also in the intriguing questions they raise about our understanding of time itself. The concept of non-linear time, often invoked in conjunction with ancient astronaut hypotheses, presents a significant departure from the linear, unidirectional model generally accepted in physics. This raises complex questions about causality, chronology, and the very nature of reality.

If we entertain the possibility of non-linear time, the interpretation of archaeological evidence undergoes a radical transformation. The seemingly impossible precision of ancient structures, for example, could be explained not by impossibly advanced technology within a confined temporal frame, but by the influence of a civilization operating outside the constraints of our linear temporal experience. This civilization might have interacted with humanity at various points throughout history, leaving behind artifacts and knowledge seemingly out of sync with the conventional chronological framework. This approach allows for a reconceptualization of technological advancements, suggesting that human history may be punctuated by episodes of highly advanced technological development interspersed with periods of relative stagnation, a pattern not readily explained by a strictly linear model of progress.

Consider the perplexing case of the Antikythera Mechanism, an intricate ancient Greek device for calculating astronomical positions. Its complexity and sophistication, discovered to far exceed the technological capabilities attributed to its era, have led to speculation regarding its origin. While advancements in our understanding of ancient Greek engineering are providing increasingly plausible explanations for its creation within the historical context, the possibility of non-linear temporal influences remains a tantalizing, albeit highly speculative, alternative. If a more advanced civilization, operating outside our linear timeline, had interacted with ancient Greek society, even indirectly, it might account for this seemingly anachronistic technological leap. The mechanism might not represent a point of indigenous technological innovation, but rather a transfer of knowledge or technology across time, defying the limitations of conventional linear chronology.

Similarly, the advanced mathematical and astronomical knowledge embedded in various ancient cultures could be interpreted through the prism of non-linear time. The precision of astronomical observations in ancient civilizations, like the Maya, who demonstrated an astonishing grasp of celestial mechanics, might suggest contact with a civilization possessing significantly more advanced knowledge of the cosmos. This knowledge, transmitted across a non-linear timeline, could have profoundly shaped the development of these cultures, accounting for their exceptional insights into the workings of the universe. The seemingly abrupt appearance and disappearance of highly advanced civilizations, evident in the archaeological record, could also be explained by temporal discontinuities, suggesting that these societies may have existed outside of, or perhaps in multiple overlapping, temporal frameworks.

However, the concept of non-linear time presents considerable challenges. Einstein's theory of relativity revolutionized our understanding of space and time, demonstrating their interconnectedness and demonstrating that time is not absolute but is relative to the observer and affected by gravity. However, even relativity doesn't accommodate the radical departures from linearity proposed in some interpretations of non-linear time. The paradoxes inherent in time travel, for instance, are well-known, and the potential for disrupting causality presents a significant obstacle to accepting the idea of non-linear temporal interference in human history. The notion of a civilization interacting with our own across vast stretches of non-linear time raises questions about the preservation of information across such temporal shifts and the mechanisms by which such interaction could even occur.

The exploration of non-linear time brings us into the realm of theoretical physics, potentially involving concepts like wormholes, parallel universes, and higher dimensions. These are areas of intense scientific debate and speculation, with few concrete empirical findings to support the claims. String theory, for instance, proposes the existence of extra spatial dimensions, which, if true, could fundamentally alter our understanding of space-time. But even within these theoretical frameworks, the precise mechanics of temporal manipulation remain largely unknown.

Moreover, the intertwining of ancient astronaut theories and non-linear time inevitably involves interpreting ambiguous evidence. The subjective interpretation of ancient texts and symbols is exacerbated by the introduction of non-linear time, as the meaning and context of such artifacts can be further obscured by the possibility of temporal distortion. What appears to be a technological anomaly from a linear perspective might be a coherent element within a non-linear temporal framework, but without a solid understanding of the principles governing this non-linearity, such interpretations remain speculative at best.

Furthermore, the lack of direct evidence for non-linear temporal effects hinders the scientific validation of theories incorporating this concept. While some argue that unexplained archaeological anomalies constitute indirect evidence, this is insufficient to establish a causal connection with non-linear time. To establish the scientific validity of non-linear time, direct empirical evidence is required, for example, repeatable and verifiable experimental results demonstrating temporal anomalies. Currently, there is a significant lack of such evidence, making claims of non-linear temporal influence highly speculative.

To reconcile ancient astronaut theories with non-linear time would require a revolutionary shift in our understanding of physics and cosmology. It would demand not only evidence for the existence of advanced extraterrestrial civilizations but also empirical proof of non-linear temporal phenomena. While the notion is intellectually stimulating, it currently lacks the robust scientific foundation to be considered a credible explanation for archaeological or historical anomalies.

The challenge, therefore, lies not in dismissing the possibility of non-linear time outright, but in establishing a rigorous framework for investigating its potential. This requires a multidisciplinary approach, integrating insights from physics, cosmology, archaeology, and even philosophy, to develop testable hypotheses and design experiments that could potentially provide empirical evidence supporting or refuting the concept. Until such evidence emerges, the connection between ancient astronaut theories and non-linear time must remain within the realm of fascinating speculation, rather than firmly established scientific fact. The allure of such possibilities should drive further scientific exploration, but it is crucial to maintain a critical and skeptical approach, avoiding the pitfalls of confirmation bias and unsubstantiated claims.

In conclusion, while the incorporation of non-linear time into ancient astronaut theories adds an intriguing layer of complexity, it simultaneously increases the speculative nature of these hypotheses. The absence of empirical evidence for non-linear temporal phenomena, coupled with the methodological challenges of interpreting ambiguous evidence, significantly limits the scientific credibility of this combined framework. While the exploration of such concepts pushes the boundaries of our understanding, rigorous scientific methodology demands that extraordinary claims require extraordinary evidence, a standard that currently remains unmet. The journey toward unraveling the mysteries of the past should be guided by a commitment to sound scientific principles, acknowledging both the limits of our current knowledge and the potential of future discoveries.

The preceding discussion explored the tantalizing intersection of ancient astronaut theories and the concept of non-linear time. While intellectually stimulating, this combination faces significant criticism, primarily stemming from a lack of empirical evidence and methodological challenges in interpreting ambiguous data. Addressing these criticisms requires a careful examination of the counterarguments and a balanced assessment of the strengths and weaknesses of the proposed framework.

One of the most significant criticisms levied against ancient astronaut theories, regardless of the temporal framework applied, is the reliance on circumstantial evidence and subjective interpretations. Proponents often point to architectural marvels, sophisticated astronomical knowledge, and unexplained artifacts as proof of extraterrestrial intervention. However, critics argue that these phenomena can be explained through alternative hypotheses grounded in conventional historical and scientific understanding. For example, the precision of ancient structures like the pyramids at Giza, often cited as evidence of advanced technology beyond the capabilities of ancient civilizations, has been increasingly attributed to sophisticated engineering techniques, advanced knowledge of mathematics and geometry, and the use of simple tools over extended periods, rather than the intervention of extraterrestrials.

The interpretation of ancient texts and symbols also presents a considerable challenge. Proponents often interpret ambiguous passages or symbolic imagery as evidence of contact with extraterrestrial beings or as instructions on advanced technologies imparted by extraterrestrial visitors. However, such interpretations are highly subjective and often lack consensus among scholars. Multiple interpretations are often equally plausible, and the selection of a specific interpretation often reflects the pre-existing beliefs and biases of the interpreter, raising concerns about confirmation bias. Without a rigorous methodology for interpreting these ambiguous sources, relying on them as evidence for advanced extraterrestrial technology or intervention remains highly problematic.

Furthermore, the concept of non-linear time, while intriguing, introduces further complexity and ambiguity. Even if we were to entertain the possibility of non-linear temporal phenomena, the mechanisms by which such phenomena would interact with human history remain largely unknown. The lack of a well-defined theoretical framework makes it difficult to formulate testable hypotheses or to design experiments capable of validating or refuting the claims. While theoretical physics explores concepts such as wormholes and parallel universes, these remain largely theoretical, with no empirical evidence to support their existence or their capacity to influence the past or present.

The counterargument to the criticism of a lack of empirical evidence is often the suggestion that the very nature of non-linear time makes direct observation challenging. It is argued that events occurring outside of our linear temporal framework may leave behind subtle traces, but not necessarily easily detectable ones. However, this lack of readily available evidence doesn't validate the hypothesis. The burden of proof lies on those proposing the theory to provide at least some form of credible evidence, even if indirect, that supports the proposition of non-linear time interaction. Without any demonstrable link between the supposed extraterrestrial involvement and the observed phenomena, the theory remains fundamentally speculative.

Another significant criticism focuses on the potential for anachronistic interpretations. The discovery of technologies or knowledge in ancient cultures that appear "out of place" within their chronological context are often presented as evidence of advanced extraterrestrial interaction. However, the constant advancement of archaeological and historical research frequently reveals new insights, offering alternative explanations for seemingly anachronistic artifacts or knowledge. Archaeological discoveries constantly reshape our understanding of ancient civilizations and technologies. Therefore, attributing anomalies to extraterrestrial interaction prematurely dismisses the potential for future discoveries to provide alternative, more conventional explanations.

The counterargument might be that the rate of discovery and our improved understanding of ancient capabilities are themselves influenced by non-linear factors, such as the very extraterrestrial intervention we are discussing. This circular reasoning, however, only strengthens the need for a clear delineation between speculative theories and established facts. Without demonstrating a verifiable connection between non-linear time and observable phenomena, the argument remains unfalsifiable, a fundamental weakness in any scientific theory.

A further criticism arises from the potential for misinterpreting technological limitations. Ancient societies, though lacking modern technology, possessed sophisticated skills and knowledge. The absence of technologies we now take for granted does not automatically imply a lack of advanced knowledge or engineering skills. Proponents of ancient astronaut theories often underestimate the resourcefulness and ingenuity of ancient peoples. Instead of seeking extraterrestrial explanations for every technological advancement, a deeper exploration of the capabilities and knowledge of these ancient cultures is crucial.

Furthermore, the application of Occam's Razor – the principle that the simplest explanation is usually the best – is frequently ignored in ancient astronaut theories. While the intervention of an advanced extraterrestrial civilization interacting with humanity across non-linear time may seem a compelling narrative, it is significantly more complex than alternative explanations that rely solely on human ingenuity and natural processes. Until more straightforward explanations are exhaustively explored and ruled out, resorting to extraordinary claims of extraterrestrial intervention and non-linear time remains scientifically premature.

The discussion of ancient astronaut theories and non-linear time should not be viewed as a dismissal of the intriguing possibilities offered by these concepts. Instead, it's a call for a rigorous and scientific approach to investigation. This requires a multi-disciplinary collaboration between archaeologists, historians, physicists, and other relevant specialists to develop methodologies for investigating potential evidence, devising testable hypotheses, and evaluating the validity of different interpretations. A critical and skeptical attitude should be paramount, constantly questioning assumptions and actively seeking to refute the proposed theories rather than merely seeking confirmation.

In conclusion, the intersection of ancient astronaut theories and non-linear time presents numerous fascinating avenues for intellectual exploration. However, the lack of empirical evidence, the methodological challenges associated with interpreting ambiguous evidence, and the potential for subjective interpretations significantly undermine the scientific credibility of the combined framework. While the possibility of extraterrestrial visitation and non-linear temporal phenomena remains a tantalizing prospect, it is crucial to approach such claims with a healthy dose of skepticism and a commitment to rigorous scientific methodology. The extraordinary claims of advanced civilizations manipulating time require extraordinary evidence—a standard that remains currently unmet. The pursuit of knowledge should be driven by a desire to understand the world through rigorous investigation, not by a pre-conceived notion of what the answers should be. The pursuit of truth demands an honest and critical evaluation of the evidence, even if it challenges our most exciting and imaginative theories.

The previous section examined the criticisms leveled against the convergence of ancient astronaut theories and non-linear time, highlighting the lack of empirical evidence and the methodological challenges inherent in interpreting ambiguous data. However, the exploration of these seemingly outlandish concepts shouldn't be dismissed entirely. A crucial aspect of this investigation lies in the analysis of ancient myths and legends—narratives that often contain elements suggestive of advanced technologies, celestial visitors, and temporal anomalies that defy linear time. These stories, passed down through generations, may offer clues, however veiled, to the possibility of non-linear temporal phenomena and extraterrestrial interaction.

One compelling area of investigation centers on recurring motifs found in myths across vastly different cultures. Consider the widespread prevalence of flood myths, found in nearly every corner of the globe—from the Epic of Gilgamesh in Mesopotamia to the biblical story of Noah's Ark and the Hindu tale of Manu. While these narratives might be interpreted as reflections of actual historical events, such as catastrophic floods, they could also be interpreted symbolically, as representing a disruption or resetting of time, a cyclical pattern of creation and destruction that transcends our linear understanding of history. The "great deluge" could represent a disruption in the flow of time, a catastrophic event that fundamentally altered the course of human civilization, perhaps even wiping out previous, more technologically advanced civilizations whose remnants remain hidden from us in a layer of our non-linear timeline. This possibility, while highly speculative, allows us to view these myths through a different lens.

Another intriguing aspect of ancient mythology is the frequent depiction of beings with superhuman abilities, often described as gods or demi-gods. These figures possess powers that seem to defy the laws of physics as we understand them, including abilities that could be interpreted as manifestations of temporal manipulation. In numerous cultures, myths feature gods who can travel across vast distances instantaneously, seemingly disregarding the limitations of space and time. Such accounts, whether literal or metaphorical, suggest a deeper understanding of the universe, potentially hinting at a form of time travel or manipulation of the temporal fabric. These narratives could serve as metaphorical representations of advanced scientific principles or technologies, inaccessible to us due to our limited present understanding but potentially explainable through the lens of quantum physics. Consider the possibility of some ancients having stumbled upon advanced technologies, possibly through extraterrestrial contact, long lost to the ravages of time and only hinted at through their surviving mythology.

The depiction of cyclical time in ancient cosmologies also deserves consideration. Many ancient cultures conceived of time not as a linear progression, but as a cyclical process of creation, destruction, and renewal. This cyclical view of time is fundamentally different from the linear model that dominates modern Western thought. This cyclical perspective could reflect a deeper understanding of temporal dynamics, perhaps even an intuitive awareness of non-linear phenomena. The Maya calendar, with its intricate cycles and predictions, provides a striking example of this cyclical worldview. The Mayan understanding of time, far more complex than our simplistic linear view, could be reflecting a more accurate, though possibly non-linear, understanding of universal processes. It remains possible that the Mayans, or other similarly advanced ancient cultures, interacted with external sources of information, allowing them to develop a deeper and more comprehensive understanding of time and universal cycles. Their advanced calendar system could even be considered a form of rudimentary time travel prediction, in the sense that it potentially predicted patterns of non-linear time's influence on humanity.

Focusing specifically on the possibility of extraterrestrial interaction, we can examine myths that feature descriptions of beings from the sky, often associated with advanced technologies and supernatural abilities. Many ancient cultures have legends of flying machines or chariots of the gods, capable of traversing vast distances. These narratives might be interpreted as metaphorical descriptions, or they might reflect encounters with advanced extraterrestrial civilizations possessing technologies far beyond our present capabilities. These technologies, especially if operating on non-linear temporal principles, could lead to accounts of strange and almost supernatural occurrences, as witnessed by the ancient peoples. Their tales, passed down through oral tradition and then into writing, could contain echoes of their experiences with beings from beyond our comprehension. One can consider the Vedic texts of India, containing descriptions of advanced flying vehicles that seem to defy our modern understanding of technology. These detailed descriptions, coupled with the other unusual features of the texts, warrant a more thorough investigation, especially when considering the possibility of non-linear interactions that could have influenced both the technology's existence and the preservation of the accounts.

Analyzing the consistency of such myths across different cultures is another vital aspect of this study. The presence of common motifs and narratives in disparate societies, geographically and chronologically separated, could suggest a shared source of information, potentially pointing towards a widespread interaction with a non-human intelligence, or possibly a universal pattern of non-linear time's influence. If non-linear time is a factor, some accounts would be more consistent than others across different cultures, given that some events would manifest within a shared, overlapping segment of non-linear time. This would need to be investigated to determine the consistency of the narrative across cultures and to assess the potential influence of non-linear factors.

The interpretation of these myths requires a careful and nuanced approach, acknowledging the limitations of the source material and the potential for misinterpretations. The symbolic nature of many myths makes direct translation challenging. However, by approaching these narratives with an open mind and a critical perspective, we can potentially uncover insights that challenge our linear perception of time and the limitations of our current understanding of history and the universe. The very act of cross-referencing these accounts across geographically separated cultures could provide a validation for the reality of non-linear time's influence. The presence of consistent, yet unusual, themes across these accounts might suggest an underlying universal mechanism at work, such as a manipulation of time itself, beyond the capabilities of ordinary human comprehension or intervention.

Furthermore, the potential for unconscious or unintentional transmission of information across cultures shouldn't be discounted. The theory of memes, initially proposed by Richard Dawkins, suggests that ideas can propagate and evolve across cultures much like genes. A shared experience, whether real or imagined, might spread through memetic transmission, leading to similar myths arising independently in different societies. These shared myths could have been influenced by a shared experience of non-linear time or extraterrestrial interaction, imprinted on the collective unconscious of humanity. Their consistent nature could be the result of a single, powerful event reverberating through time, albeit across a non-linear temporal spectrum.

The investigation of ancient myths and legends, therefore, isn't about proving the existence of ancient astronauts or non-linear time but rather about exploring the possibility and considering the implications of such concepts. It's about broadening our perspective and challenging our assumptions about the past, present, and future, acknowledging that our understanding of time and history is incomplete, and the possibility remains that our understanding of reality is just as incomplete. The study of ancient myths, approached rigorously and with an open yet critical mind, could potentially uncover valuable insights that challenge our understanding of the universe and humanity's place within it. The pursuit of knowledge necessitates the constant questioning of our assumptions, and these ancient myths provide an intriguing and potentially enlightening avenue for exploration. The consistent presence of unusual accounts and themes across various cultures might reveal more than just the imaginative storytelling of our ancestors. It might potentially reveal the fragmented and elusive truth of a complex reality extending beyond our current linear understanding of time.

5 Non-Linear Time and the Human Experience

Building upon the analysis of ancient myths and their potential allusions to non-linear time, we now turn our attention to a more contemporary, and arguably more accessible, avenue of exploration: the human experience of anomalous temporal phenomena. Specifically, we will examine déjà vu and precognitive experiences, events that, while subjective and often dismissed as mere quirks of memory or coincidence, might offer compelling insights into the possibility of non-linear temporal interactions.

Déjà vu, the unsettling feeling of having already lived through a present moment, is a remarkably common experience. Estimates suggest that the vast majority of adults will experience déjà vu at some point in their lives. The sheer prevalence of this phenomenon argues against its dismissal as an isolated neurological anomaly. While various neurological explanations have been proposed—ranging from minor temporal lobe seizures to simple memory glitches—these explanations often fail to fully account for the vividness and intensely subjective nature of the experience. The feeling isn't simply one of familiarity; it's a profound sense of having already witnessed the unfolding events, complete with accompanying sensory details and emotional responses. This intensity suggests something more profound than a simple malfunction of the memory system.

One intriguing hypothesis proposes that déjà vu represents a brief, fleeting glimpse into a parallel timeline or a future iteration of the present moment. In a non-linear temporal model, multiple timelines might coexist, and brief "cross-talk" or interferences between these timelines could manifest as déjà vu. This "cross-talk" may not be a conscious experience; it might occur at a subconscious level, manifesting as a feeling of familiarity rather than a complete recall of a past event. This theory aligns with the concept of "parallel universes" or "multiverse" models explored in modern physics, suggesting that our subjective experience of time might be only one facet of a much more complex temporal reality.

Further supporting this hypothesis is the frequent association of déjà vu with heightened emotional states or moments of intense focus. These states might lower the threshold for such temporal bleed-through, allowing glimpses of alternate timelines to intrude into our conscious experience. The intensity of the emotion, the concentration of focus, or perhaps even the particular energetic state of the individual, could act as a catalyst for this temporal interference. Alternatively, the act of remembering itself might create a ripple in the fabric of time, drawing forth a reflection from another point in the non-linear timeline. While highly speculative, this perspective allows us to view déjà vu not as a neurological error, but as a subtle interaction between different temporal strata.

Precognitive experiences, on the other hand, represent an even more radical challenge to our linear understanding of time. These experiences involve having knowledge of future events before they occur, often with a remarkable level of detail and accuracy. While often attributed to coincidence or confirmation bias—the tendency to remember instances that confirm our beliefs and ignore those that don't—many documented cases defy such simplistic explanations. Anecdotal accounts, often recounted in personal journals or testimonies, describe future events predicted with startling accuracy, involving specific details that could not have been reasonably anticipated.

From a non-linear perspective, precognitive experiences could be interpreted as glimpses into possible futures, or even inevitable futures, existing within a broader temporal landscape. These glimpses might not be predictions in the traditional sense, but rather observations from a vantage point outside the confines of our linear experience of time. The brain, according to this theory, might be capable of accessing information from different points in this non-linear temporal stream, receiving perceptions of events that are yet to occur within our conventional timeline.

The mechanisms underlying such precognition remain largely unknown, but several hypotheses have been proposed. Some researchers propose a connection to quantum entanglement, suggesting that interconnectedness at a quantum level could allow for information transmission across time. Others have explored the possibility of a collective unconscious, a reservoir of shared psychic energy that encompasses all points in time, allowing for an intuitive apprehension of future events. Still others propose a form of temporal resonance, whereby certain individuals may be more sensitive to vibrations emanating from the future, receiving these impressions as precognitive insights. These hypotheses, while theoretical, underscore the profound implications of precognitive experiences in relation to non-linear time.

The investigation of these phenomena necessitates a multidisciplinary approach. Neurological studies could help identify the brain structures and processes involved in both déjà vu and precognitive experiences. Psychological studies could shed light on the subjective experience and interpretative biases associated with these events. And finally, a more open-minded exploration of the concepts within the realm of parapsychology and quantum physics might uncover underlying mechanisms that defy conventional understanding. One possibility, for example, is that certain individuals possess heightened sensitivity to temporal fluctuations, enabling them to experience non-linear temporal interactions more frequently. The study of such individuals might provide valuable insights into the nature of time itself and the capacity of the human consciousness to interact with it in unexpected ways.

It's crucial to acknowledge the skepticism surrounding these topics. The lack of conclusive empirical evidence and the difficulty in designing rigorous experiments have hampered scientific investigation. However, the persistent recurrence of these anomalous experiences and their cultural ubiquity, across various cultures and historical periods, necessitates continued exploration. To dismiss them outright is to ignore a potential window into the very nature of time and the limits of our current scientific understanding.

Moreover, the very act of experiencing déjà vu or precognition might alter the fabric of non-linear time itself. The act of observing the future, even fleetingly, might introduce a feedback loop, subtly influencing the trajectory of events. This concept aligns with the observer effect in quantum physics, suggesting that the act of observation itself can change the outcome of an experiment. Similarly, the experience of déjà vu might represent a subtle interaction with an alternate timeline, the very act of experiencing this interference altering the original timeline's course. This could explain why the events perceived in a déjà vu experience sometimes seem to deviate slightly from expectations, and why a sense of foreboding sometimes accompanies precognitive experiences. The consciousness itself may be an active participant in the weaving of time, rather than merely a passive observer.

Furthermore, the interpretation of these experiences needs to be approached with caution, avoiding confirmation bias and other cognitive pitfalls. Detailed and rigorous documentation of each instance, paying attention to the specific details, the circumstances surrounding each experience, and the subsequent unfolding of events, is crucial for ensuring the reliability and credibility of these cases. This attention to detail might uncover previously overlooked patterns, highlighting the interplay between conscious experience, memory, and the potentially non-linear nature of time.

In conclusion, while the study of déjà vu and precognitive experiences remains a challenging and somewhat controversial area of inquiry, these phenomena offer intriguing avenues to explore the possibility of non-linear time. By integrating a multidisciplinary approach and acknowledging the limitations and challenges, we can move forward towards a more holistic understanding of human experience and the universe. The seemingly inexplicable nature of these experiences pushes the boundaries of our current understanding, inviting a reassessment of our fundamental assumptions about the nature of time and the capabilities of the human consciousness. These seemingly paradoxical experiences may not be mere glitches in our perception but rather windows into a more complex reality, a reality that transcends our conventional linear understanding of time and space. The ongoing investigation of these phenomena will undoubtedly continue to challenge, provoke, and ultimately reshape our comprehension of the universe and our place within it.

The exploration of déjà vu and precognitive experiences provides a fertile ground for considering the nature of time and consciousness. However, to fully grasp the potential implications of a non-linear temporal framework, we must broaden our scope to encompass other anomalous human experiences: out-of-body experiences (OBEs) and near-death experiences (NDEs). These phenomena, often relegated to the fringes of scientific inquiry, offer compelling evidence that challenges our conventional understanding of personal identity, spatial location, and the very nature of time itself.

Out-of-body experiences, characterized by the sensation of being separated from one's physical body, while seemingly subjective and easily dismissed as hallucinations or psychological distortions, often involve a remarkable sense of clarity and heightened awareness. Individuals reporting OBEs frequently describe a panoramic view of their surroundings, observing their physical body from an external perspective, often with an extraordinary level of detail. This detail extends beyond mere visual perception; they frequently report hearing conversations, feeling sensations, and even experiencing emotional states within their environment, all from their seemingly disembodied viewpoint. Furthermore, many OBE reports include veridical details—details that could not be known through ordinary sensory perception—further complicating purely psychological explanations.

The implications of veridical perceptions during OBEs are profound. If an individual can accurately perceive their surroundings while ostensibly outside their physical body, it suggests that consciousness is not solely confined to the brain but might exist independently, perhaps even non-locally. This challenges the materialist view that equates consciousness with brain activity, proposing instead a more expansive model where consciousness is a non-local phenomenon, not bound by the constraints of physical space and potentially even time.

Connecting OBEs to a non-linear temporal framework requires us to consider the possibility that consciousness might not be confined to a single point in space-time but might traverse different temporal points, or even exist simultaneously across multiple temporal coordinates. In this model, the OBE isn't merely a departure from the physical body but a shift in the individual's consciousness' primary point of reference in the spacetime continuum. The individual's conscious awareness might momentarily detach from the physical body, "traveling" to another point in space-time, perceiving events from that altered perspective before returning to its original location in the body. This would explain the often-reported experiences of accessing information seemingly unavailable through normal sensory means, or even the perception of events occurring in the future or past.

Near-death experiences (NDEs), often occurring during moments of extreme physiological stress such as cardiac arrest or severe trauma, share remarkable similarities with OBEs. Individuals reporting NDEs frequently describe leaving their body, observing their surroundings, and undergoing profound spiritual or transcendental experiences. Common themes include sensations of peace and tranquility, feelings of floating or flying, encounters with deceased loved ones, and a review of one's life.

Like OBEs, many NDE accounts contain veridical perceptions—details that could only be known to someone observing the scene from outside the physical body. These reports raise fundamental questions about the nature of consciousness and its relationship to the body. Are these experiences simply hallucinations caused by oxygen deprivation or other physiological stresses? Or do they provide evidence for the existence of a non-local consciousness that can transcend the limitations of the physical body and potentially experience time non-linearly?

The concept of non-locality, a cornerstone of quantum physics, further strengthens the connection between OBEs, NDEs, and non-linear time. Quantum entanglement demonstrates that two or more particles can become linked in such a way that their fates are intertwined, regardless of the distance separating them. A measurement performed on one particle instantaneously affects the other, suggesting a connection that transcends spatial limitations. This quantum entanglement might, theoretically, extend beyond the subatomic level, encompassing consciousness itself.

If consciousness is non-local, it might exist outside the constraints of space-time, potentially accessing information from different points in time or across multiple temporal dimensions. This would provide a framework for understanding OBEs and NDEs not as neurological anomalies but as manifestations of consciousness's non-local nature, allowing for interactions with non-linear time.

Furthermore, the temporal aspects of OBEs and NDEs are often described as highly subjective and potentially non-linear. Many accounts describe experiences of time dilation—the sensation that time is either speeding up or slowing down drastically. Time during these events often feels compressed or expanded, contradicting our conventional linear perception of time. These subjective temporal shifts might be indicative of the consciousness navigating a non-linear temporal landscape, experiencing events outside the framework of our usual linear perception.

Another significant aspect linking OBEs and NDEs to non-linearity is the frequent reporting of profound insights and altered perspectives on life and reality. These life-changing shifts in perspective, often accompanied by a deep sense of spiritual awareness, could potentially stem from an exposure to a different temporal framework—a kind of broader cosmic perspective that transcends the limitations of our linear, egocentric view of time and existence. These altered states of awareness, following such experiences, are rarely temporary, often leading to significant life changes in the individual's values, beliefs, and behaviors, highlighting the potential for a profound and lasting impact on one's temporal and existential understanding.

The scientific investigation of OBEs and NDEs faces significant challenges. The subjective nature of these experiences makes them difficult to quantify and measure objectively. Moreover, ethical concerns surrounding inducing such experiences limit the possibility of controlled experiments. However, the sheer number of documented accounts, their consistency across cultures and historical periods, and the frequently reported veridical perceptions warrant further investigation.

The rigorous scientific investigation of OBEs and NDEs requires a multidisciplinary approach, integrating elements of neuroscience, psychology, parapsychology, and quantum physics. Neurological studies could focus on identifying the brain states and physiological mechanisms associated with these experiences, exploring whether specific brain regions or processes are involved. Psychological studies could delve into the subjective experience itself, exploring the cognitive processes that shape the interpretation of these events, considering potential biases and the influence of cultural narratives.

The inclusion of parapsychology brings in the investigation of psychic abilities and potential interactions with non-local or non-linear realms. This exploration, while often met with skepticism, opens the door to exploring possible mechanisms that might mediate information transfer across time and space. Finally, the principles of quantum physics could provide a theoretical framework for understanding the non-locality of consciousness and its potential interactions with a non-linear spacetime continuum.

The study of OBEs and NDEs, while challenging and often controversial, offers unique opportunities to expand our understanding of consciousness, time, and the nature of reality itself. By adopting a multidisciplinary approach and remaining open to the possibility of non-linear temporal interactions, we can advance our knowledge of the human experience and its relation to the cosmos. Dismissing these phenomena as mere hallucinations or psychological anomalies is to neglect a potential window into a reality far richer and more complex than our current understanding allows. The continuing exploration of these experiences might fundamentally reshape our comprehension of our existence and our place within the vast, unfolding tapestry of time.

Our exploration of non-linear time and its potential manifestations in human experience thus far has focused on out-of-body experiences (OBEs) and near-death experiences (NDEs). These phenomena, while challenging to study scientifically, suggest a potential for consciousness to interact with time in ways that defy our conventional linear model. However, another intriguing avenue for investigating non-linear temporality lies in the realm of seemingly random events that, upon closer examination, reveal a surprising interconnectedness: synchronicity.

Carl Jung, the renowned Swiss psychiatrist and psychoanalyst, coined the term "synchronicity" to describe "meaningful coincidences"—events that seem causally unrelated but are connected through their psychological or symbolic significance. These are not mere coincidences in the statistical sense; instead, they possess a quality of "meaningfulness" for the individual experiencing them. For instance, an individual deeply contemplating a specific problem might repeatedly encounter symbols or images related to that problem throughout their day—in books, conversations, or even seemingly random visual encounters. This isn't a matter of confirmation bias; the synchronistic events often feel strikingly relevant and deeply resonant, impacting the individual's emotional and cognitive state. The sheer improbability of these coincidences occurring randomly further strengthens the argument for an underlying connection.

Jung posited that synchronicity represents a manifestation of acausal connections—connections that are not governed by the principles of cause and effect as understood in classical physics. He proposed that synchronicity points to a deeper, underlying order in reality, one that operates beyond the limitations of linear, deterministic processes. This "underlying order" could be interpreted as an interconnectedness within a non-linear temporal framework, where seemingly disparate events are linked through a hidden, non-causal mechanism. The experience of synchronicity suggests that certain events might be connected not through linear causality, but through a deeper, perhaps non-temporal, interconnectedness.

Consider, for example, the common experience of thinking about someone and then receiving a call or message from that person shortly afterward. From a purely statistical perspective, such an event might not be improbable, especially given the volume of communication in modern life. However, the psychological impact of such events is undeniable; the feeling of a profound connection, the sense that something more than mere chance is at play, strongly suggests a non-random occurrence. This isn't merely about the statistical likelihood; it's about the felt experience, the resonance between the inner thought and the external event. This resonance transcends the confines of conventional causality.

The significance of synchronicity extends beyond individual experiences. The collective experience of seemingly synchronized events, such as widespread feelings of anxiety or anticipation preceding major historical events, offers another intriguing dimension to the exploration of non-linear temporal connections. These phenomena, while difficult to quantify scientifically, raise questions about the potential for a collective unconscious, a shared mental field influencing events across space and time. The potential influence of such a field could indicate a deeper level of interconnectedness in human experience, transcending individual consciousness and implying a non-local interaction with the flow of time.

One might argue that these are merely instances of confirmation bias—a tendency to notice and remember events that confirm pre-existing beliefs. While confirmation bias undeniably plays a role in the interpretation of personal experience, it doesn't fully explain the sheer improbability of many synchronistic events, nor the depth of feeling associated with them. These experiences consistently transcend the individual's expectations and, frequently, possess a powerful transformative effect.

From a non-linear temporal perspective, synchronicity could be interpreted as a manifestation of events being interconnected across different temporal coordinates. The seemingly random event occurring in the present might be linked to a past event or even a future event through a non-linear, acausal pathway. This pathway transcends our conventional understanding of cause and effect, implying a deeper, more holistic understanding of time. Instead of a linear progression, time could be perceived as a multi-dimensional landscape, with events interconnected through non-linear pathways.

The scientific investigation of synchronicity presents considerable challenges. The subjective nature of the experience makes it difficult to quantify and measure objectively. Moreover, the very definition of "meaningful coincidence" is culturally and personally influenced, making universal criteria for assessment elusive. The lack of a universally accepted model also hinders scientific rigor. However, the universality of these experiences across cultures and historical periods, and the consistent feeling of profound interconnectedness experienced by individuals, suggest a phenomenon worthy of exploration.

A rigorous approach to studying synchronicity would necessitate a multidisciplinary effort. Psychology would explore the subjective experiences and cognitive biases influencing interpretation. Parapsychology could investigate possible links to non-local consciousness and psi phenomena, such as telepathy or precognition, suggesting a potential pathway for acausal connections. While these connections are controversial, the possibility of interconnectedness between minds and the temporal flow, through a shared, non-local consciousness field, cannot be ignored. Finally, theoretical physics, particularly exploring quantum entanglement and non-local phenomena, could provide a theoretical framework for understanding acausal connections within a non-linear temporal model.

The study of synchronicity, while challenging, offers a unique window into the nature of time and consciousness. It allows us to move beyond the constraints of linear causality and consider a more holistic, interconnected model of reality. To disregard these experiences as mere chance occurrences is to miss a potential clue to a more profound understanding of the human experience and our place within the universe. The ongoing exploration of synchronicity, coupled with the ongoing examination of OBEs and NDEs, could fundamentally reshape our perception of time and our comprehension of the intricate interconnectedness of all things, expanding our awareness beyond the limitations of our current linear perspective. The possibility that we are interwoven within a rich, non-linear tapestry of time, influencing and being influenced by events across various temporal coordinates, is a captivating and significant possibility for further investigation. The ongoing study of this phenomenon, though fraught with challenges, may unveil profound insights into the fabric of reality itself.

The integration of insights from various fields—from quantum physics and parapsychology to cognitive science and philosophy—is crucial for a comprehensive understanding of synchronicity. Furthermore, the integration of these insights with our understanding of OBEs and NDEs creates a powerful framework for exploring the nature of consciousness and its interaction with time. These seemingly disparate fields of study converge in their challenge to the conventional linear model of time and their implication of a more profound and interconnected reality, one where consciousness is not confined to a single point in spacetime but might interact with, influence, or potentially even traverse, different temporal coordinates. The journey to unlock these mysteries requires open minds, a multidisciplinary approach, and a willingness to confront the limitations of our current paradigms. The rewards, however, promise to be transformative to our understanding of our place within the universe and the vast, intricate tapestry of time itself.

Building upon our exploration of synchronicity and its potential implications for a non-linear understanding of time, we now turn our attention to the fascinating realm of intuition and its possible connection to precognition. Intuition, that elusive inner knowing that seems to bypass the usual processes of rational thought, has captivated philosophers, psychologists, and even physicists for centuries. Often described as a "gut feeling" or a sudden flash of insight, intuition presents a compelling challenge to our conventional understanding of cognition and information processing.

The cognitive processes underlying intuition remain a subject of intense debate. Some researchers propose that intuition represents the rapid processing of vast amounts of information accumulated unconsciously, allowing for seemingly instantaneous judgments and decisions. This unconscious processing model suggests that our brains are constantly sifting through information, forming patterns and making connections beyond our conscious awareness. When a situation arises that resonates with these pre-existing patterns, an intuitive response surfaces, appearing as a spontaneous insight.

This unconscious processing might not be limited to information gathered linearly through time. If time itself is non-linear, as the preceding chapters have suggested, then perhaps our unconscious mind can access information from different temporal coordinates. This would imply that intuition isn't merely the result of past experiences being rapidly synthesized, but a potential avenue for accessing information from the future or even parallel timelines.

Neurologically, the exact mechanisms underlying intuition remain poorly understood. However, research suggests the involvement of several brain regions, including the anterior cingulate cortex, which is implicated in conflict monitoring and decision-making, and the insula, associated with interoception—our awareness of our internal bodily states. Activity in these areas often precedes conscious awareness of an intuitive insight, suggesting that unconscious processing occurs prior to conscious experience.

Furthermore, studies involving brain imaging techniques like fMRI have revealed that intuitive judgments are often accompanied by changes in brainwave patterns, specifically in the alpha and theta frequency bands, which are associated with relaxed states of consciousness and heightened suggestibility. This could indicate that a state of altered consciousness is conducive to accessing non-linear temporal information, opening up pathways beyond the usual limitations of linear thinking.

However, even if intuition can be understood as a form of rapid unconscious processing, the possibility of precognition—knowledge of future events—presents a far more radical challenge to our understanding of time and causality. Precognition, often associated with psychic abilities or paranormal phenomena, suggests that consciousness can access information that has yet to occur in linear time. This is not simply predicting the future based on current trends or probabilities, but rather a form of direct access to events yet to unfold.

Several research methodologies have attempted to investigate precognitive abilities. These include experiments involving Ganzfeld procedures, remote viewing, and precognitive dream studies. While these studies have generated some interesting results, they have also been fraught with challenges, including methodological limitations, susceptibility to bias, and the difficulty in replicating findings across different laboratories. The inherent subjectivity of precognitive experiences also makes objective measurement incredibly difficult.

However, the persistence of anecdotal accounts and the subjective experience of precognition cannot be dismissed entirely. The sheer number of individuals reporting precognitive experiences suggests a phenomenon worthy of deeper investigation, even if we lack rigorous scientific confirmation. If these experiences are to be understood in the context of non-linear time, they point to a profound interconnectedness of events, transcending the conventional limitations of linear cause and effect.

Consider the well-documented phenomenon of "déjà vu"—that unsettling feeling of having already experienced a current situation. While psychological explanations exist for déjà vu (such as memory glitches or neurological misfirings), its pervasiveness suggests a possible link to temporal anomalies. Perhaps déjà vu represents a brief glimpse into a non-linear temporal pathway, a fleeting intersection with a parallel timeline or a future event that momentarily resonates with the present.

The possibility of accessing non-linear temporal information, whether through intuition or precognition, raises profound implications for our understanding of consciousness. It suggests that consciousness is not merely a product of brain activity confined to a specific point in spacetime, but a more fundamental phenomenon capable of interacting with time in unconventional ways.

This perspective aligns with some interpretations of quantum physics, where the concept of superposition suggests that particles can exist in multiple states simultaneously. If consciousness is fundamentally linked to quantum processes, as some researchers suggest, it could imply that our consciousness might exist, at least partially, in multiple temporal states, accessing information across different temporal coordinates.

Furthermore, the study of intuition and precognition could provide a valuable bridge between science and spirituality. Many spiritual traditions posit the existence of a non-linear temporal framework, where past, present, and future exist simultaneously. These traditions often describe intuitive insights and prophetic visions as forms of accessing information from this unified temporal field. A rigorous scientific approach to these phenomena, rather than dismissing them out of hand, could potentially lead to a more holistic and comprehensive understanding of consciousness and reality.

Moreover, the exploration of these areas is not solely an intellectual exercise. Understanding the potential for accessing non-linear temporal information has profound implications for human potential. If we can improve our ability to access intuition and even enhance our precognitive capabilities (however unlikely that might seem), we could potentially make more informed decisions, anticipate future challenges, and enhance our creative problem-solving abilities.

The challenges in researching intuition and precognition are considerable. The subjective nature of the experiences, the difficulty in establishing objective criteria, and the potential for bias make rigorous scientific investigation exceptionally complex. However, the potential rewards of a deeper understanding of these phenomena, and their implications for our understanding of time, consciousness, and reality itself, are immense. A multidisciplinary approach, encompassing cognitive science, neuroscience, parapsychology, and theoretical physics, is crucial for making progress in this field.

Ultimately, the exploration of intuition and precognition within a non-linear temporal framework is a journey into the unknown. It requires a willingness to question our assumptions about the nature of time and causality, to embrace the limitations of our current scientific understanding, and to approach the subject with open minds and a healthy dose of intellectual humility. The journey will undoubtedly be challenging, but the potential insights offer a glimpse into a profound and transformative understanding of ourselves, our place in the universe, and the very fabric of time itself. The potential for unlocking previously unimagined dimensions of human experience, based upon a more complete understanding of time's non-linear properties, is a compelling reason to continue this vital research. This exploration extends far beyond the simple realm of psychic abilities; it delves into the very essence of consciousness and its relationship to the universe, a quest that warrants dedicated and rigorous investigation. Only through continued inquiry and open-minded investigation can we hope to uncover the secrets held within the seemingly paradoxical realm of intuition and precognition, and thereby illuminate our understanding of the intricate and profound nature of time itself.

The exploration of non-linear time, as we have been undertaking, inevitably leads us to consider the subjective human experience. While scientific models and theoretical frameworks provide a crucial foundation for understanding, the lived experiences of individuals who claim to have encountered temporal anomalies offer a rich, albeit less quantifiable, source of insight. These personal accounts, though often lacking the rigor of controlled experiments, can provide valuable clues and provoke new avenues of inquiry. It is crucial, however, to approach these narratives with a critical yet open mind, acknowledging the potential for subjective biases and the limitations of anecdotal evidence.

One common theme emerging from these personal accounts is the experience of "déjà vu," that unsettling feeling of familiarity with a novel situation. While neurological explanations involving memory glitches or misfirings are often cited, the sheer frequency of déjà vu suggests a more profound phenomenon might be at play. Several individuals describe experiences that extend beyond a simple feeling of familiarity. One account details an individual who, during a business meeting in a previously unvisited city, experienced a vivid sense of having lived through that specific conversation, down to the smallest details of body language and subtle nuances of speech. This was not merely a vague sense of recognition, but a profound and unsettling feeling of having relived the moment before. Such intensely detailed accounts challenge purely neurological interpretations and suggest a possible interaction with a parallel timeline or an echo from a future event.

Another recurring theme in personal accounts centers on precognitive dreams or intuitions. These range from vague feelings of impending events to strikingly accurate predictions. Numerous individuals report dreaming of events that later occurred, sometimes with startling specificity. One frequently cited example involves a dream of a plane crash that accurately predicted the details of a subsequent real-world event, including the aircraft type, the time of the crash, and even the location of the wreckage. While these accounts are often dismissed as coincidences, the cumulative weight of such similar testimonies warrants further consideration. The possibility that these dreams reflect access to non-linear temporal information demands a re-evaluation of our understanding of dream mechanisms and their potential relationship to temporal processes. Scientifically, these accounts challenge the notion of linear causality, posing the question: if a dream can accurately reflect a future event, does it imply a form of causal influence across time? This could indicate a subtle but potentially significant interaction between the dreamer's subconscious and future events.

Moreover, personal accounts involving synchronicity, particularly those involving significant coincidences or meaningful events occurring in close temporal proximity, suggest a possible underlying non-linear temporal mechanism. One particularly striking account involves an individual who, after years of searching, found a long-lost relative through a series of seemingly improbable coincidences, each event independently unlikely, yet occurring in sequence, leading to a reunion thought highly improbable. These experiences, although subjectively interpreted, highlight the potential for events to align in a manner that defies purely probabilistic explanations, suggesting a deeper level of interconnection that might be explained through a non-linear model of time.

In contrast to instances of precognitive dreams or heightened intuition, some accounts describe experiences of temporal distortion. These are less about accessing specific information from other times and more about a subjective alteration in the perceived flow of time. Individuals describe moments where time seems to accelerate or decelerate, or where periods of time feel compressed or stretched. Such subjective experiences may reflect the fluidity of time perception, but they also raise the possibility of an actual non-linear temporal experience. For instance, an individual who recounts a car accident where time seemed to slow down dramatically could potentially have experienced an alteration of temporal flow, rather than simply an effect of heightened alertness or stress.

It's crucial to distinguish between subjective perception and objective reality. Many of these experiences can be explained through psychological factors, such as stress, memory biases, or cognitive distortions. However, the sheer volume and consistency of certain patterns within these accounts warrant further investigation. The prevalence of specific themes, such as precognitive dreams or striking instances of synchronicity, cannot be simply dismissed as isolated occurrences. Instead, these patterns might hint at underlying mechanisms that operate beyond our current understanding of time and consciousness. A holistic approach, which integrates the subjective accounts with neuroscientific data and theoretical physics, is essential to develop a comprehensive understanding.

The challenge lies in bridging the gap between anecdotal evidence and rigorous scientific investigation. Gathering reliable, verifiable data from personal accounts is difficult. The inherent subjectivity of experience necessitates rigorous methodologies that account for biases and potential misinterpretations. Researchers might consider utilizing structured questionnaires, employing standardized methods for assessing the quality and consistency of reported experiences, and conducting qualitative analysis to identify recurring patterns. This structured approach could help to filter out anecdotal accounts dominated by biases and emotional interpretations and focus on instances containing more detailed, repeatable elements.

Furthermore, incorporating advanced neuroimaging techniques could shed light on the neural correlates of non-linear temporal experiences. By studying brain activity patterns during these experiences, we might uncover specific neural signatures associated with altered temporal perception, precognition, or heightened intuition. Correlating subjective reports with objective physiological data is crucial for strengthening the scientific validity of these accounts and for further understanding of the potential relationship between the neural substrate and the lived experience of temporal anomalies.

The analysis of personal accounts is not simply about validating or refuting claims of extraordinary experiences. Rather, it serves as a valuable tool for identifying areas that require further scientific scrutiny and to help refine our theoretical models. By understanding the common themes and consistent patterns emerging from these narratives, we can better design experiments to test hypotheses related to non-linear time and consciousness. In essence, the subjective accounts can inspire the objective investigation, directing the focus of scientific inquiry towards potentially fertile research avenues. The goal is not to prove or disprove the reality of non-linear time, but to refine our understanding of its possible manifestations in human consciousness, employing a methodology that embraces both the subjective and objective approaches. This integrated approach, while fraught with difficulties, promises to be far more fruitful than either approach in isolation. The task ahead is immense, requiring a multidisciplinary approach that respects the rigor of scientific inquiry while maintaining an open-minded approach to the complexities and the profound mystery of human consciousness in relation to the flow of time.

6 Non-Linear Time in Popular Culture

The enduring appeal of "Doctor Who," a science fiction series spanning decades, lies partly in its playful yet profound engagement with the concept of non-linear time. Unlike many time travel narratives that grapple with paradoxes as plot devices, "Doctor Who" uses them as a springboard for philosophical and existential explorations. The Doctor, a time-traveling alien, traverses millennia, encountering countless civilizations and witnessing the unfolding (and sometimes unraveling) of history. This inherently non-linear approach to storytelling allows the show to address complex themes of causality, free will, and the nature of identity in ways that resonate deeply with viewers.

One of the most compelling aspects of "Doctor Who's" time travel mechanics is its apparent disregard for the "grandfather paradox," a classic conundrum that arises when altering the past potentially prevents one's own existence. While the show occasionally alludes to temporal disturbances and potential consequences of altering the timeline, it rarely dwells on the logical inconsistencies. Instead, it prioritizes narrative dynamism and character development, often suggesting that the universe, or the Doctor's own enigmatic abilities, somehow accommodates these seemingly impossible actions. This creative license allows for compelling stories that wouldn't be possible within a rigidly adhered-to framework of linear causality. The Doctor's encounters often involve rescuing individuals or civilizations from impending doom, actions that inevitably change the course of history. Yet, rather than exploring the potential ramifications of these changes, the focus remains on the immediate consequences and the emotional impact on the characters involved.

The show's depiction of alternate timelines and parallel universes further complicates the linear model of time. Episodes frequently feature alternate versions of the Doctor, companions, and even entire planets, each existing within its own unique temporal stream. This exploration of multiplicity challenges the singular, fixed narrative often associated with linear time, suggesting that reality itself is a multifaceted, ever-evolving entity. The concept of fixed points in time, events that cannot be altered regardless of the Doctor's intervention, also adds another layer of complexity. These fixed points act as anchors within the seemingly fluid fabric of time, suggesting an underlying structure despite the apparent chaos of temporal manipulation.

Furthermore, "Doctor Who" cleverly utilizes the concept of time as a resource. Time, in the show, isn't just a backdrop; it's a commodity to be managed, wasted, or treasured. The Doctor's travels are often driven by the urgency of preventing catastrophic events, emphasizing the value of time and the responsibility that comes with the ability to manipulate it. This concept resonates with our own experience, as we constantly grapple with the finite nature of our time and the choices we make within its constraints. The show's depiction of time as both precious and plentiful simultaneously offers a unique perspective on our temporal limitations, reminding us of both the urgency and the potential contained within each passing moment.

Beyond the narrative structure, the show's visual aesthetic also contributes to its unique portrayal of non-linear time. The often fantastical and surreal environments that the Doctor visits reflect the fluidity and unpredictability of temporal journeys. The seemingly impossible juxtapositions of past, present, and future further reinforce this sense of temporal disorientation, making the viewer question the stability and linearity of their own experience of time. The use of anachronisms, the intentional blending of different historical periods, adds another layer to this effect, showcasing the show's deliberate playfulness with the constraints of temporal order.

The companions, who often represent a more grounded, human perspective, serve as a crucial counterpoint to the Doctor's time-bending adventures. Their reactions to these extraordinary events highlight the contrast between the Doctor's perceived mastery of time and the human struggle to comprehend its complexities. Their responses – ranging from awe and excitement to fear and confusion – resonate with the viewers' own potential reactions to similar scenarios, strengthening the engagement and facilitating an exploration of non-linear time through a relatable human lens.

However, the show's treatment of time isn't without its critics. Some argue that its casual approach to paradoxes undermines the internal consistency of the narrative. The lack of stringent adherence to the rules of causality can lead to plot inconsistencies and a sense of narrative arbitrariness. This perceived lack of rigor, however, can be seen as a deliberate stylistic choice, prioritizing engaging storytelling over strict adherence to scientific principles. The show's strength lies not in its scientific accuracy but in its imaginative exploration of the possibilities that arise from a non-linear model of time.

"Doctor Who" frequently uses time travel as a means of exploring moral and ethical dilemmas. The Doctor is often forced to make difficult choices, weighing the potential consequences of interfering in past events against the imperative to prevent future catastrophes. These moral quandaries highlight the complexities of temporal manipulation, forcing viewers to confront the potential ramifications of their own actions, regardless of their temporal context. The show subtly encourages reflection on our responsibilities in shaping the future and the limitations of our understanding of long-term consequences.

The character of the Doctor himself embodies the complexities of non-linear time. The Doctor's multiple regenerations, transformations into entirely new bodies, suggest a fluid, non-linear relationship with identity. This ability to transcend the confines of a single physical form underscores the theme of continuous change and evolution, highlighting the continuous process of becoming that extends beyond the linear progression often associated with aging. The Doctor's experiences across time, the accumulation of knowledge and wisdom, further solidify this concept of a non-linear personal journey, contrasting starkly with the linearity often associated with human aging.

"Doctor Who's" consistent popularity, across generations and cultural contexts, speaks to its success in translating complex scientific concepts into accessible and engaging narratives. The show's ability to seamlessly blend science fiction with philosophical reflection makes it a unique and enduring exploration of non-linear time in popular culture. It invites viewers to question their assumptions about causality, identity, and the nature of reality itself, reminding us that time, while often perceived as linear, might be far more fluid and complex than we typically assume. The show ultimately demonstrates that the exploration of non-linear time doesn't just provide a playground for exciting adventures but also offers a rich canvas for profound philosophical reflection, inviting us to contemplate our place within the grand tapestry of time. Its enduring appeal highlights the human fascination with the nature of time and our continuous attempts to unravel its mysteries, even if, like the Doctor, we must often accept a certain degree of paradoxical ambiguity in our explorations. The show serves not as a definitive answer but as a catalyst for further exploration of the fascinating complexities of time and our own place within its ever-unfolding narrative.

The cinematic masterpiece "Interstellar" offers a compelling example of how the complexities of relativistic time dilation can be woven into a narrative, enriching its emotional impact and expanding its thematic scope. Unlike "Doctor Who's" playful disregard for temporal paradoxes, "Interstellar" grounds its narrative firmly in Einstein's theory of general relativity, using the scientific concept of gravitational time dilation as a central plot device. The film doesn't shy away from the scientific intricacies, but rather uses them to create a powerful and moving story about human resilience, sacrifice, and the enduring power of love across vast distances and timescales.

The film vividly portrays the extreme time dilation experienced by characters near a supermassive black hole called Gargantua. For every hour spent on the planet Miller, orbiting dangerously close to Gargantua, years pass on Earth. This dramatic difference in the passage of time isn't merely a plot device; it's a consequence of the immense gravitational field generated by the black hole. According to general relativity, stronger gravitational fields slow down the passage of time relative to weaker fields. This effect, while subtle in everyday life, becomes profoundly significant in the extreme gravitational environment near a black hole. The film visually represents this effect through the striking disparity between the aging of the characters on Miller's planet and those who remain orbiting further away.

The emotional weight of this scientific concept is deeply felt in the film. The characters' experiences of time dilation aren't merely intellectual exercises; they become visceral realities that dramatically shape their relationships and their understanding of their own mortality. The prolonged separation caused by the time difference creates an agonizing tension, highlighting the ephemeral nature of time and the profound impact of distance on human connection. The scenes depicting the reunion (or lack thereof) with loved ones, after years of separation dictated by relativistic effects, underscore the film's central themes of loss, sacrifice, and the enduring strength of human relationships against overwhelming odds.

"Interstellar" masterfully utilizes visual storytelling to convey the scientific concepts. The visuals showcasing the warped spacetime around Gargantua are breathtaking, mirroring the complex physics of gravitational lensing and time dilation. These are not mere special effects; they are visual representations of complex scientific ideas, making them accessible to a wider audience. The film doesn't dumb down the science, but rather uses visual metaphors and dramatic narrative to make it understandable and emotionally resonant. This blend of rigorous scientific underpinnings and compelling storytelling is a hallmark of the film's success.

Beyond the immediate dramatic tension, the film also utilizes time dilation to explore broader philosophical themes. The significant time difference between the planet Miller and Earth forces characters to confront the limitations of human lifespan and the relentless march of time. The knowledge that time flows differently for different individuals, depending on their proximity to a massive object like a black hole, challenges our conventional understanding of temporal linearity. It suggests that time isn't a universal constant, but rather a relative experience shaped by the gravitational forces of the universe.

The film's exploration of time dilation isn't simply a scientific exposition; it's a crucial element of the characters' journeys and their moral dilemmas. The choices they make are informed by their understanding of the relativistic effects of time, and the consequences of those choices are deeply intertwined with the complexities of temporal experience. For example, the decision to land on Miller's planet, despite the knowledge of the extreme time dilation, is a calculated risk with profound consequences for the crew. This decision highlights the human tendency to push boundaries, to explore even in the face of overwhelming odds, and the ethical considerations that arise in the context of limited time and immense risk.

Furthermore, "Interstellar" uses the concept of time dilation to highlight the themes of hope and perseverance in the face of seemingly insurmountable challenges. The characters' unwavering commitment to finding a new home for humanity underscores their belief in the future and their determination to overcome the limitations imposed by time and distance. The film doesn't offer easy answers or simplistic resolutions; rather, it presents a complex and nuanced portrayal of the human struggle for survival and the enduring power of hope amidst overwhelming adversity. The time dilation acts not as a barrier but as a challenge, urging characters—and the audience—to reconsider the value of time, both on a personal and a universal scale.

The film also touches upon the more philosophical implications of relativistic time. If time is relative, then the perception of simultaneity becomes problematic. Events that seem simultaneous to one observer may not be simultaneous to another, depending on their relative motion and gravitational fields. This challenges our intuitive understanding of causality, suggesting that the sequence of events might not be universally consistent. While the film doesn't explicitly delve into these philosophical intricacies, the narrative implicitly raises these questions, prompting the audience to reflect on the subjective nature of time and the limitations of our understanding of causality in the context of extreme gravitational fields.

Another important aspect is how "Interstellar" uses time as a resource. The characters are constantly battling against the clock, not just in terms of Earth's resources dwindling but also because of the ever-present threat of time dilation and the need to return to Earth before it's too late for the remaining human population. The limited time available influences every decision, shaping the narrative and adding a sense of urgency that intensifies the emotional impact of the story. This element underscores the importance of efficient resource management, not just of material resources, but also the most precious resource of all: time itself.

The film's use of non-linear storytelling techniques further reinforces its exploration of non-linear time. The narrative jumps back and forth between different time periods, mirroring the fractured experience of time caused by relativistic effects. This non-linear structure enhances the emotional impact of the story, allowing the audience to gradually unravel the complexities of the characters' journeys and the overarching plot. The film's structure becomes a reflection of the non-linear, relative nature of time itself.

Finally, "Interstellar" differs from "Doctor Who" in its scientific rigor. While "Doctor Who" embraces imaginative liberties in its depiction of time travel, "Interstellar" grounds its narrative in established scientific principles, even if it pushes the boundaries of our current understanding. This grounded approach to science fiction makes the film's portrayal of relativistic time dilation more compelling and believable, enhancing the emotional impact and the intellectual engagement of the story. The film demonstrates that scientific accuracy and imaginative storytelling need not be mutually exclusive, but rather can complement and enhance each other to create a truly profound and memorable cinematic experience. It uses science not as a crutch but as a springboard for exploration, creating a film that is both scientifically stimulating and emotionally resonant. The film succeeds because it finds the perfect balance between the rigor of scientific principles and the power of imaginative storytelling, making it a significant contribution to the representation of non-linear time in popular culture.

The playful chaos of time travel depicted in Robert Zemeckis's "Back to the Future" trilogy stands in stark contrast to the scientifically grounded approach of "Interstellar." While "Interstellar" meticulously adheres to Einsteinian physics, "Back to the Future" revels in the paradoxes and inconsistencies inherent in manipulating the timeline. This difference highlights the diverse ways popular culture engages with the concept of non-linear time, ranging from rigorous scientific exploration to whimsical flights of fancy.

The most prominent paradox explored in "Back to the Future" is the Grandfather Paradox, a classic thought experiment that questions the very fabric of causality. The paradox arises from the possibility of altering the past in such a way that one's own existence becomes impossible. If Marty McFly were to prevent his parents from meeting, for instance, he would cease to exist, thus rendering his time-traveling actions impossible in the first place. This creates a logical contradiction, a self-negating loop that challenges the linear progression of time and the presumed stability of causality.

The film cleverly sidesteps a direct confrontation with the Grandfather Paradox through a combination of narrative maneuvering and a degree of narrative ambiguity. While Marty's actions in the past do significantly alter the course of his parents' relationship, they do so in a way that ultimately ensures his own continued existence. The changes are subtle enough to avoid a complete unraveling of the timeline, allowing for a happy ending while still raising the intriguing question of how close the film came to triggering a true temporal paradox. The film's resolution doesn't provide a definitive answer to the Grandfather Paradox; it simply navigates around it, demonstrating the narrative freedom inherent in fictional representations of time travel.

Beyond the Grandfather Paradox, "Back to the Future" explores other temporal inconsistencies, such as the "multiple timelines" or "parallel universes" theory that offers a potential escape route from the paradox. The idea is that altering the past doesn't necessarily change one's own original timeline; instead, it creates an alternate timeline, branching off from the original. This allows for changes to the past without the immediate threat of self-erasure. While the film never explicitly states which theory it follows, the outcome implies an element of branching timelines, allowing Marty to return to his original, largely unchanged, future.

The film's comedic elements and lighthearted tone are key to its success in navigating such complex issues. The focus is more on the adventurous escapades and comedic mishaps than on a deep philosophical or scientific exploration of time travel. The film uses the time travel mechanism as a comedic device rather than a means of probing complex temporal dynamics. This approach allows it to sidestep the more rigorous and potentially frustrating aspects of temporal paradoxes, embracing instead a playful exploration of what such a phenomenon *might* be like, without the need for absolute consistency. The film's entertainment value lies not in the meticulous adherence to temporal rules, but in the inventive ways it uses them as a catalyst for comedic and dramatic developments.

This contrasts sharply with the seriousness with which "Interstellar" handles time dilation. "Back to the Future" employs time travel as a means of creating humorous situations and exciting adventures, whereas "Interstellar" uses it to explore the emotional and psychological consequences of relativistic time. The films represent two different approaches to storytelling, reflecting the vast spectrum of artistic interpretations of time in popular culture.

Furthermore, "Back to the Future" employs a form of temporal "tinkering," where minor adjustments to the past are made to achieve desired outcomes. These are less about major historical shifts and more about personal interventions. This approach avoids the potentially chaotic consequences of altering major historical events, and it focuses more on the consequences for Marty's family rather than the whole world. This focus on personal consequences keeps the scale of the narrative manageable and allows for a more clearly defined resolution, avoiding the complexities of potentially world-altering changes.

The film's use of time travel isn't merely a plot device; it functions as a commentary on the importance of family, personal responsibility, and the inherent value of one's own life. Marty's journey isn't just about fixing the past; it's about understanding his own place within the larger framework of his family and his community. This personal narrative makes the exploration of temporal paradoxes more relatable, grounding the fantastical elements within the familiar context of human relationships.

The sequels further elaborate on the possibilities and paradoxes of time travel, introducing alternative timelines and the concept of unforeseen consequences. These sequels show how even minor changes in the past can have ripple effects throughout the future, showcasing the butterfly effect in a dynamic and engaging way. These sequels increase the narrative complexity and thematic depth, while still maintaining the film's comedic and adventurous tone. The recurring element of needing to "fix" a problem caused by time travel is a recurring theme, emphasizing the responsibility that comes with such power.

The fact that "Back to the Future" remains a cultural icon decades after its release attests to its effectiveness in navigating complex ideas in an engaging way. Its success lies not in providing definitive answers to the paradoxes of time travel, but in exploring these paradoxes through a lens of humor, adventure, and personal growth. It embraces the inherent ambiguity and uncertainty of altering the past, while still maintaining narrative coherence and emotional satisfaction.

In contrast to the scientific accuracy sought in "Interstellar," "Back to the Future" prioritizes narrative consistency within its own internally consistent rules, even if those rules defy scientific accuracy. This allows for creative freedom and greater exploration of the story's core themes. It allows for a lighthearted approach to otherwise mind-bending concepts, making it accessible and enjoyable to a wider audience. This approach highlights the potential of using non-linear time not just as a scientific puzzle, but as a storytelling tool capable of exploring themes of personal identity, family relationships, and the consequences of choices across time.

The film's legacy lies in its ability to engage audiences with the fascinating and potentially disturbing implications of time travel, without overwhelming them with scientific jargon or philosophical complexities. It uses these concepts to fuel a highly engaging and entertaining narrative, leaving the audience to grapple with the implications long after the credits roll. The film's impact extends beyond its entertainment value, prompting reflection on the nature of time and its role in shaping our lives. Its enduring popularity demonstrates the enduring human fascination with non-linear time and its potential for both comedic and dramatic effect. This makes "Back to the Future" a significant example of how popular culture can explore the complex and fascinating world of non-linear time, without necessarily sacrificing entertainment for scientific accuracy. The film shows that the imaginative exploration of non-linear time can be both entertaining and thought-provoking, a testament to the power of creative storytelling. The very act of time travel becomes a metaphor for the transformative power of our actions, emphasizing that even seemingly small choices can have profound and far-reaching effects. The film's popularity shows the widespread human fascination with the idea of controlling or altering the past, a reflection of our own desires for control and second chances.

The Butterfly Effect, a cornerstone of chaos theory, posits that even the smallest change in initial conditions can have drastically large and unforeseen consequences over time. This concept, inherently linked to non-linearity, finds fertile ground in narratives exploring time travel and alternate realities. While "Back to the Future" playfully dances around the direct implications of altering the past, other works delve more explicitly into the chaotic ripples stemming from seemingly insignificant actions.

Consider Ray Bradbury's chilling short story, "A Sound of Thunder," where a time-traveling safari inadvertently alters the future by stepping on a butterfly. This seemingly minor act unleashes a cascade of changes, transforming the language, political landscape, and even the very appearance of the future. Bradbury masterfully illustrates the fragility of the timeline and the terrifying potential for unforeseen consequences inherent in tampering with the past. The story's power lies not in intricate scientific explanations, but in the stark and visceral depiction of the butterfly effect's ramifications. The subtle shift in the present, the seemingly insignificant act of crushing an insect, has monstrous consequences in the future, making it a powerful allegory for the responsibility inherent in our actions, however seemingly small. The altered future is not simply a different version of the original timeline; it is fundamentally different, highlighting the pervasive and insidious nature of the butterfly effect.

The butterfly effect's narrative potential extends beyond outright time travel. In many stories, the effect is implied, woven into the fabric of the narrative without explicitly invoking time machines or temporal paradoxes. Consider the intricate plots of detective novels, where a seemingly inconsequential clue, a missed phone call, or a fleeting glance can dramatically shift the course of an investigation, leading to entirely different outcomes. The chain reaction of events, subtly influenced by minor details, echoes the butterfly effect, emphasizing how seemingly insignificant choices can shape destinies. The very structure of many mystery novels relies on the unpredictable consequences of seemingly small actions, reflecting the chaotic nature of the butterfly effect in a subtle yet effective way. The reader is often kept in suspense, not knowing the full extent of the consequences of a single action until much later in the story, mirroring the unpredictable nature of the butterfly effect itself.

The impact of the butterfly effect is not limited to literature and speculative fiction. Consider the realm of historical counterfactuals, where historians explore "what if" scenarios, examining how seemingly minor events might have altered the course of history. What if Archduke Franz Ferdinand had not been assassinated? What if the Roman Empire had adopted Christianity earlier? These explorations, though speculative, highlight the butterfly effect's relevance beyond the realm of fantasy. They show that pivotal moments in history were not inevitable, but were instead the product of a complex interplay of circumstances, demonstrating how even seemingly small events, such as a delayed train or a sudden change of plans, could have dramatically shifted the course of empires and civilizations. The impact of these historical what-ifs showcases how history is not a linear progression, but a complex, chaotic system deeply influenced by the butterfly effect.

Film often employs the butterfly effect to dramatic effect. In the film "Sliding Doors," the protagonist's life unfolds in two parallel narratives, based on whether or not she catches a train. This simple choice leads to drastically different life paths, showcasing the potentially life-altering implications of seemingly insignificant decisions. The film cleverly uses the narrative structure to highlight the dramatic consequences of small actions, illustrating the butterfly effect without explicit reference to time travel. Each path she takes highlights the impact of choice, demonstrating the extent to which life is shaped by even seemingly trivial occurrences. The narrative structure itself mirrors the chaotic, unpredictable nature of the butterfly effect, showing the multiple possibilities that could unfold from a single choice.

The butterfly effect also permeates the narrative structure of many science fiction films, but often in a less explicit manner than in "A Sound of Thunder." Films like "Donnie Darko," while complex and not solely centered on the butterfly effect, hint at the chaotic nature of time and the interconnectedness of events. The film's nonlinear narrative structure, its cryptic symbols and dream sequences, all contribute to a sense of inevitability intertwined with moments of chaotic contingency. The seemingly random events of the film are linked through a complex web of causality, where seemingly small actions have significant consequences, reflecting the essence of the butterfly effect. The film's ambiguity and reliance on interpretation make it a powerful expression of the butterfly effect's uncertainty and inherent unpredictability, adding another layer to its complex narrative.

Even in films that don't explicitly focus on time travel, the butterfly effect can serve as a thematic undercurrent. Consider character-driven dramas where a single impulsive act, a moment of weakness, or a missed opportunity sets off a chain of events with far-reaching and often tragic consequences. The protagonist's choices, often made in ignorance of their ultimate consequences, underscore the concept of unintended repercussions and the profound impact even seemingly minor decisions can have. The narrative arc often highlights the unintended consequences of actions, serving as a cautionary tale about the butterfly effect's pervasive influence on human lives. These films show that the butterfly effect is not confined to theoretical physics or speculative fiction, but rather is a powerful metaphor for the unpredictable nature of human life and the often unforeseen consequences of our choices.

In examining the butterfly effect across popular culture, it becomes evident that its appeal lies in its capacity to highlight the interconnectedness of events and the fragility of causality. It's a concept that resonates deeply because it speaks to the inherent uncertainty of life, the unpredictable nature of consequence, and the profound impact of seemingly insignificant choices. The butterfly effect isn't just a scientific principle; it's a potent narrative device that adds depth, complexity, and a sense of dramatic irony to stories, illustrating the unpredictable consequences of actions and the often-unforeseen ramifications of human choices. The fascination with the butterfly effect in film and literature is a reflection of our own struggle to understand and control the seemingly chaotic unfolding of life itself. The narrative interpretations of the butterfly effect also highlight a deeper understanding of personal responsibility and the consequences of choices across both linear and non-linear time. The concept serves as a potent metaphor for the intricate web of cause and effect that shapes both individual lives and the broader sweep of history.

The exploration of non-linear time in popular culture wouldn't be complete without a thorough examination of time loops, a narrative structure that has captivated audiences for decades. Unlike the butterfly effect, which focuses on the cascading consequences of a single action, time loops present a cyclical structure where events repeat, often with variations driven by the protagonist's actions within the loop. This cyclical structure provides a unique framework for exploring themes of fate, free will, and the nature of reality itself. The inherent limitations and possibilities within the confines of the loop offer fertile ground for complex character development and compelling narratives.

One of the most prominent examples of a time loop narrative is "Groundhog Day," a film that brilliantly uses the repetitive structure to explore themes of personal growth and redemption. Phil Connors, a cynical weatherman, is trapped in a time loop, reliving the same day repeatedly. Initially, he uses this newfound power for self-serving purposes, indulging in hedonistic pleasures. However, as the loop continues, he begins to confront his own flaws and ultimately undergoes a profound transformation, learning empathy and compassion. The film's success lies not only in its comedic timing but also in its poignant exploration of the human condition, using the time loop as a mechanism for character development and thematic exploration. The repetitive nature of the loop allows for a deep dive into Phil's psyche, revealing his inner struggles and eventual redemption. The ending, where he breaks free from the loop, is both satisfying and deeply meaningful, illustrating the transformative power of self-improvement and the possibility of escaping predetermined fate.

The time loop structure allows for a unique type of narrative tension. Unlike linear narratives that build tension through anticipation of future events, time loops create suspense by focusing on the protagonist's attempts to manipulate the loop itself. Each iteration of the loop builds upon the previous one, with the protagonist learning from past mistakes and attempting to alter the outcome. The audience is kept in suspense, wondering whether the protagonist will ever escape the loop or find a way to alter the trajectory of events. This creates a different kind of suspense, not about an unknown future, but about the possibility of success within a known, albeit repetitive, framework. This suspense isn't about what *will* happen, but what the protagonist *can* do to change what *does* happen.

"Edge of Tomorrow," a science fiction film featuring Tom Cruise, offers a more action-oriented take on the time loop narrative. Here, the loop is tied to a specific battle against alien invaders, with the protagonist reliving the same day repeatedly until he manages to find a way to defeat the enemy. The film uses the time loop as a device to explore themes of combat, strategy, and sacrifice. Unlike "Groundhog Day," the focus is less on character development and more on the strategic possibilities presented by the loop. The protagonist uses each iteration to gain experience, refine his combat skills, and learn the enemy's tactics. The film deftly balances action sequences with strategic planning, offering a unique blend of action and suspense. The tension stems not just from the loop itself, but also from the escalating stakes of the battles and the ever-present threat of death within the confines of the repeating day.

Beyond film, time loops are also employed effectively in literature and video games. Many video games, particularly those with rogue-like elements, incorporate time loops into their gameplay mechanics. Players repeatedly face similar challenges, learning from their mistakes and adapting their strategies with each iteration. This creates a unique gameplay experience that rewards persistence and careful observation. The repetitive nature of these games also allows for a deeper understanding of the game's mechanics and strategy. The frustration of repeating failure is balanced by the thrill of eventual success and mastery of the game's systems, creating a compelling and engaging gameplay loop. The challenge becomes not just about winning, but about achieving mastery through repeated failure and adjustment.

Furthermore, the concept of time loops frequently emerges in literary works, often intertwined with elements of magical realism or science fiction. These narratives frequently use time loops to explore existential themes such as the meaning of life, the nature of free will, and the cyclical nature of time itself. Authors may employ the time loop to create a framework for examining the consequences of choices and actions, highlighting the impact of even seemingly minor decisions within the context of a repeating temporal structure. The limited scope of the loop allows for a focused exploration of these themes, providing the author with a clear and concise platform for exploring complex philosophical questions. The inherent constraints of the loop often become a metaphor for the limitations and possibilities of human existence itself.

The narrative possibilities of time loops extend beyond simple repetition. Some works incorporate variations within the loop, allowing for subtle changes in the environment or the characters' interactions, hinting at the potential for influence within a seemingly fixed framework. This creates a dynamic tension where the protagonist's actions gradually alter the trajectory of the loop, suggesting that even within constraints, there's a possibility of genuine change and progress. The subtle variations within the repetition prevent the narrative from becoming monotonous, introducing elements of surprise and uncertainty that keep the audience engaged. This subtle deviation keeps the audience guessing about the eventual outcome, creating suspense that is distinct from other forms of storytelling.

Moreover, the psychological impact of being trapped in a time loop is a rich area for exploration. Many narratives delve into the mental and emotional toll of reliving the same period repeatedly, showcasing how this could impact a person's sense of self, their relationships with others, and their perception of reality. This provides an opportunity to examine the nature of identity and the resilience of the human psyche under extreme pressure. The emotional arc of a character trapped in a time loop is often far more complex and nuanced than in linear narratives, creating characters with profound depth and empathy. The challenges faced within the loop are not simply physical; they are also deeply psychological.

In conclusion, time loops offer a potent narrative tool for exploring a wide range of themes, from personal growth and redemption to strategic combat and existential questions about the nature of reality. The cyclical structure provides a unique framework for building suspense, developing characters, and examining complex philosophical ideas. Their use in film, literature, and video games highlights the enduring appeal of this non-linear narrative structure and its versatility in expressing various aspects of the human condition and our understanding of time itself. The adaptability of the time loop structure also allows for creative variations and inventive storytelling, suggesting that this narrative device will continue to find applications in various forms of media in the years to come, offering an ever-evolving landscape of narrative possibilities and profound engagement with the human experience.

7 Philosophical Implications of Non-Linear Time

The exploration of non-linear time, particularly as depicted in narratives like time loops, significantly impacts the age-old philosophical debate surrounding free will versus determinism. Traditional determinism posits a universe governed by immutable laws of cause and effect, where every event is a pre-ordained consequence of prior events. In this view, free will is an illusion; our choices are merely the inevitable outcome of a chain of prior causes stretching back to the universe's inception. However, the very concept of a non-linear temporal structure, as presented in time loops, offers a potent counterargument to this rigid deterministic perspective.

If time isn't strictly linear, if events can loop, repeat, and be influenced by actions within the loop itself, then the deterministic chain of cause and effect is disrupted. The existence of a time loop implies a degree of temporal flexibility, a break in the otherwise rigid sequence of cause and effect that underpins traditional determinism. Within the confines of the loop, the protagonist's actions demonstrably affect the subsequent iterations. This inherent malleability suggests an element of agency, a potential for genuine choice and influence that challenges the core tenets of determinism.

Consider the example of "Groundhog Day." Phil Connors doesn't simply passively relive the same day; he actively participates in shaping his experience. His initial self-serving actions lead to negative consequences, reinforcing the loop. However, as he progressively chooses to act differently, to demonstrate empathy and compassion, the loop begins to change. This implies that his choices, however situated within the loop, possess genuine causal power; they are not simply predetermined reactions to prior events. His eventual escape from the loop, a direct consequence of his altered behavior, strongly suggests a capacity for genuine free will, a defiance of any predetermined fate.

The implications extend beyond fictional narratives. Quantum physics, with its probabilistic nature and inherent uncertainty principle, provides a scientific basis for challenging strict determinism. The observation of a quantum system inherently influences its state, demonstrating that the act of observation itself participates in shaping reality. This parallels the interactive nature of time loops; the act of experiencing the loop, and acting within it, alters the loop's trajectory. While not a direct equivalence, both phenomena challenge the notion of a perfectly predictable, deterministic universe.

However, the counterarguments to this interpretation are significant. Some might argue that even within a time loop, the loop itself is predetermined. The structure, the parameters of the repetition, and the eventual escape condition could all be fixed, pre-ordained elements of a larger deterministic framework. From this perspective, Phil Connors' choices, while appearing free within the confines of the loop, are ultimately part of a larger, predetermined narrative arc. His journey toward self-improvement is not a free choice but a necessary step in the pre-programmed unfolding of events. This perspective preserves the deterministic worldview by situating the seemingly free choices within a larger, pre-ordained context.

Furthermore, the very concept of a time loop, particularly in fictional narratives, introduces a degree of artificiality. It is a contrived narrative device, and while it allows for an exploration of free will, it doesn't necessarily prove its existence in the real world. The rules governing time loops in fiction are often arbitrarily set by the author, creating a controlled environment where the exploration of choice is simplified. This distinction between fiction and reality is crucial. What holds true in a fictional time loop may not be directly applicable to the complex, multi-layered reality we inhabit. The simplification of the temporal structure in such narratives allows for a clearer demonstration of agency, but this simplification might not reflect the complexities of real-world causality.

The debate extends to the nature of consciousness and its role in shaping reality. Some interpretations of quantum mechanics suggest that consciousness plays an active role in the collapse of the wave function, implying that our awareness actively participates in shaping reality. This resonates with the narrative structure of a time loop. The protagonist's conscious experience within the loop, their awareness of its repetitive nature, and their subsequent actions directly influence the outcome. This suggests that consciousness isn't simply a passive observer but an active participant in the shaping of reality, even within the confines of a non-linear temporal structure.

However, a purely subjective approach based on individual conscious experience also introduces significant limitations. The interpretation of a time loop's influence on free will relies heavily on the subjective perspective of the protagonist within the narrative. This subjective experience, while valuable in exploring the philosophical implications, doesn't provide a universally applicable understanding of free will. Objective criteria for determining whether actions within a time loop are truly "free" remain elusive. The very notion of objectively measuring free will is a challenge that extends beyond the confines of time loop narratives.

Another layer of complexity is introduced when we consider the potential for multiple, parallel time loops or branching narratives within a non-linear temporal framework. If multiple loops exist simultaneously, each with different outcomes determined by different choices within those respective loops, the deterministic view becomes even more problematic. The existence of multiple possibilities, rather than a singular, predetermined outcome, strongly suggests a degree of genuine choice and agency.

The possibility of branching timelines, where choices lead to diverging paths, directly challenges the single-linear causal chain of traditional determinism. Instead of a single, inevitable outcome, the universe might comprise a multitude of possible outcomes, each dependent on the choices made within each branching narrative. This creates a much more complex view of causality, one that accommodates the notion of free will within a multi-dimensional temporal framework. This understanding of time, not as a single, linear progression, but as a web of interconnecting possibilities, fundamentally alters our understanding of causality and the nature of free will.

In conclusion, the exploration of non-linear time, particularly through the lens of time loop narratives, presents a potent challenge to traditional deterministic views of the universe. While counterarguments exist, and the artificiality of many fictional portrayals needs careful consideration, the very concept of a malleable, non-linear time structure suggests the potential for genuine free will. The implications extend beyond mere narrative; they touch upon fundamental questions of causality, consciousness, and the nature of reality itself, fostering ongoing dialogue between physics, philosophy, and our understanding of human agency. The challenge lies in bridging the gap between the imaginative exploration of these concepts in fiction and the rigorous scientific and philosophical analysis required to fully grasp their implications for our understanding of the universe and our place within it. The debate continues, fueled by ever-evolving narratives and our expanding understanding of the complexities of time itself.

The exploration of non-linear time, as we've seen, profoundly impacts our understanding of reality itself. The very fabric of existence, traditionally conceived as a linear progression from past to future, becomes fluid, malleable, even potentially cyclical when viewed through the lens of non-linear temporality. This fluidity necessitates a re-evaluation of long-held philosophical positions, particularly regarding the nature of reality – the fundamental question of what is real and how we perceive it.

Two prominent philosophical standpoints, realism and idealism, offer contrasting perspectives that require re-examination in this context. Realism, in its simplest form, posits that reality exists independently of our perception of it. A tree falling in a forest makes a sound, regardless of whether anyone is there to hear it. The objective reality is paramount; our subjective experience is merely a reflection of this pre-existing, independent reality. However, if time is non-linear, if past, present, and future are interwoven, then the very concept of an independent, objectively existing reality becomes muddied. If events can influence their past, if actions in the present can retroactively alter past occurrences, as depicted in many time loop narratives, then the seemingly fixed and immutable reality of realism is challenged. The objective reality, in this case, isn't simply a pre-existing structure but rather a dynamic, ever-evolving entity constantly shaped by actions within the non-linear temporal framework.

Idealism, on the other hand, emphasizes the primacy of mind and consciousness in shaping reality. For an idealist, reality is fundamentally dependent on perception; it is, in essence, a mental construct. The implications of non-linear time align more readily with this perspective. If our actions within a non-linear temporal structure can influence past events, it suggests a powerful interplay between consciousness and the structure of reality itself. Our awareness, our conscious choices, actively participate in shaping the very fabric of time and, consequently, the reality we inhabit. The "reality" experienced within a time loop, for instance, is not merely a passive observation but a direct consequence of the protagonist's conscious engagement with the loop itself. Their choices, their actions within that loop, actively construct and reconstruct their reality within the confines of its non-linear structure.

However, even within the framework of idealism, non-linear time introduces complexities. If reality is fundamentally a mental construct, are there multiple realities coexisting simultaneously, mirroring the potential for multiple, parallel time loops? If the reality within a time loop is partially constructed by the individual's consciousness, what are the implications for the reality of others? Are their realities also being shaped and influenced by these non-linear temporal dynamics, potentially leading to divergences and inconsistencies in shared perceptions? These questions point towards a potential multiplicity of realities, a subjective experience of reality that might not always align with others. The possibility of shared versus individual experiences within a non-linear framework adds another layer of complexity to the idealistic perspective.

Beyond realism and idealism, the concept of non-linear time challenges other fundamental philosophical tenets. The nature of causality, for instance, becomes significantly more intricate. In a linear temporal framework, cause precedes effect; a clear, sequential chain of events unfolds predictably. But in non-linear time, causality can be circular, feedback loops can emerge, and the very distinction between cause and effect can become blurred. Events in the future can seemingly influence events in the past, potentially leading to paradoxical situations and challenging the traditional understanding of causal determinism. This challenges the classical view that every effect has a preceding cause in a linear sequence. The implications are far-reaching, impacting our understanding of free will, determinism, and the very predictability of the universe.

Consider the implications for scientific endeavors. Our scientific methodologies are, to a large extent, based on the assumption of linear time. Experiments are designed with a clear sequence of actions and expected outcomes, relying on the predictable progression of cause and effect. But if time is non-linear, this methodology needs to be re-evaluated. The very concept of repeatability, a cornerstone of scientific experimentation, is challenged. If the past can be altered by actions in the present, then seemingly identical experiments conducted at different times might yield different results, not due to experimental error but due to the non-linear influence of the temporal structure itself. The implications extend beyond the methodology; they challenge the very foundation of scientific understanding built upon the assumption of a linear, predictable universe.

Furthermore, the impact of non-linear time extends into the realm of ethics and morality. If actions in the present can alter the past, what are the ethical implications of such temporal manipulation? Could we retrospectively justify actions that were morally reprehensible at the time of their occurrence, simply by altering their consequences in the past? These questions point towards a potential ethical grey zone, necessitating a re-evaluation of moral frameworks in a world where time itself is not a rigid, unyielding structure.

The integration of quantum mechanics further complicates the picture. The probabilistic nature of quantum phenomena challenges the deterministic view of reality, mirroring the uncertainty inherent in non-linear time. The observer effect, where the act of observation influences the outcome of an experiment, demonstrates that consciousness plays a role in shaping reality at the quantum level. This resonates deeply with the idea of a non-linear time loop, where the conscious awareness of the protagonist actively shapes the loop's trajectory. The interplay between quantum mechanics and non-linear temporality suggests a universe that is far more fluid, interconnected, and profoundly influenced by consciousness than previously imagined.

In conclusion, the implications of non-linear time on our understanding of reality are vast and profound. It challenges fundamental philosophical perspectives, necessitates a re-evaluation of scientific methodologies, and raises intricate ethical considerations. The fluid, interconnected nature of time, as suggested by non-linear models, fundamentally alters our view of causality, free will, the nature of reality, and our place within the universe. The journey of understanding these implications is ongoing, and its destination remains uncertain, a testament to the ever-evolving nature of knowledge and our capacity for exploring the mysteries of time itself. This exploration necessitates a cross-disciplinary approach, weaving together physics, philosophy, psychology, and potentially even parapsychology to forge a more holistic understanding of this enigmatic dimension of existence. The potential for such an interdisciplinary approach is exciting, offering the possibility of unprecedented insights into the nature of reality and the human condition. The ongoing debate will undoubtedly refine our comprehension of time, reality, and our place within the vast cosmic tapestry.

The inherent linearity of cause and effect, a cornerstone of classical physics and philosophy, crumbles under the weight of non-linear time. In a linear temporal framework, causality operates as a straightforward chain: A causes B, B causes C, and so on. This predictable sequence allows for the formulation of scientific laws, the prediction of future events based on current observations, and the establishment of a seemingly deterministic universe. Newtonian physics, for example, thrives on this linear causality, elegantly describing the predictable interactions of objects in motion. However, the introduction of non-linear time throws this established paradigm into disarray.

If the future can influence the past, as suggested by certain interpretations of non-linear temporal models, the very concept of a singular, unidirectional causal chain becomes untenable. Imagine a scenario where an event in the future (C) retroactively alters a prior event (A), thereby influencing the event that supposedly caused it (B). This creates a closed loop, a feedback mechanism where causality becomes circular, rather than linear. Such a scenario directly challenges the fundamental principle that cause must precede effect.

This circular causality is not merely a theoretical curiosity; it finds echoes in various areas of human experience. Consider the phenomenon of self-fulfilling prophecies. A belief in a future outcome (C) – perhaps a belief in one's impending failure – can unconsciously shape present actions (B), ultimately leading to the realization of that very belief (A). The belief in the future, therefore, retroactively influenced the past actions that resulted in the dreaded future. This doesn't negate the reality of the actions themselves, but it highlights how the future's perceived influence can fundamentally shape the present and, consequently, the past.

The implications extend beyond individual psychology. Consider historical events. The outcome of a battle (C) might retrospectively alter the strategic decisions made before the battle (B), even if those decisions were seemingly independent of the future result at the time they were made. The narrative of history itself can be reinterpreted through the lens of non-linear time, questioning the deterministic view of historical progression as an inevitable sequence of cause and effect.

Furthermore, the problem of causality within a non-linear temporal framework is exacerbated by the potential for multiple timelines or parallel universes. If the manipulation of the past generates branching timelines, then the concept of a singular, overarching causal chain becomes even more problematic. Each branch, each alternate reality, might possess its own unique causal network, with events potentially having different causes and effects depending on the specific timeline. This challenges the notion of a universally applicable causal principle, suggesting instead that causality itself is context-dependent and relative to the specific temporal trajectory.

The impact on scientific methodology is profound. The scientific method relies heavily on the predictability of cause and effect. Experiments are designed to isolate variables, control conditions, and observe the consistent effects resulting from specific causes. Repeatability is crucial for validating scientific findings. However, in a non-linear temporal system, the same experiment conducted at different times might yield different results, not because of experimental error, but because of the inherent instability of causality within a non-linear framework.

This doesn't necessarily invalidate the scientific method, but it calls for a critical re-evaluation and adaptation. It might necessitate a shift towards probabilistic models, acknowledging the inherent uncertainty and unpredictability inherent in a non-linear system. Scientists may need to incorporate considerations of temporal context, acknowledging that the same cause may not always produce the same effect due to temporal fluctuations and influences from the future. This requires a more holistic approach, incorporating elements of complexity theory and possibly even embracing the uncertainties inherent in quantum mechanics.

Quantum mechanics itself introduces further challenges to the classical understanding of causality. The probabilistic nature of quantum events challenges the deterministic view of reality. The Heisenberg Uncertainty Principle, for instance, indicates that we cannot simultaneously know both the position and momentum of a particle with perfect accuracy. This inherent uncertainty mirrors the unpredictability inherent in non-linear time, where future events can retroactively alter past actions, defying a strict deterministic causal chain.

The observer effect in quantum mechanics further complicates the picture. The act of observing a quantum system inevitably alters the system itself, highlighting the inextricable link between the observer and the observed. This resonates with the concept of non-linear time, where conscious awareness might play a significant role in shaping the temporal trajectory. A conscious observer within a time loop, for instance, actively participates in shaping the loop's progression, demonstrating the potent interplay between consciousness and the causal structure of reality.

The philosophical implications are far-reaching. The concept of free will, for example, is profoundly challenged. If future events can influence past actions, to what extent are we truly free in our choices? Are our actions predetermined by future influences, rendering the concept of free will illusory? Or is free will not simply the ability to act freely, but also the ability to consciously navigate and possibly alter these non-linear causal loops?

Furthermore, the implications for ethical considerations are profound. If the past can be altered, does this open the door to retroactive justification of morally reprehensible actions? Could a negative consequence be negated by retroactively changing the actions that led to it? The very foundations of moral responsibility might need to be re-evaluated within a non-linear temporal context. Such considerations necessitate a significant philosophical shift, potentially challenging traditional conceptions of justice and accountability.

In conclusion, the problem of causality within the framework of non-linear time necessitates a radical rethinking of our understanding of the universe. It challenges the deterministic, linear models that have dominated scientific and philosophical thought for centuries. The fluid, interconnected nature of causality within non-linear time calls for a more holistic, integrated approach, incorporating insights from various disciplines – physics, philosophy, psychology, and possibly even parapsychology – to fully grapple with the complexities of this enigmatic temporal dimension. The implications are vast and profound, affecting not only our understanding of the universe, but also our understanding of ourselves and our place within the cosmic order. The journey to a comprehensive understanding of non-linear causality is ongoing, requiring a willingness to embrace uncertainty and challenge long-held assumptions about the nature of time, reality, and the very fabric of existence itself. The exploration of these complex questions continues to push the boundaries of human knowledge and understanding.

The exploration of non-linear time inevitably leads us to a crucial intersection: the nature of consciousness. If the past, present, and future are not rigidly sequential, but instead interact in complex, interwoven ways, how does this affect our understanding of subjective experience? The classical view, rooted in linear causality, portrays consciousness as a product of the brain, a phenomenon arising from the intricate electrochemical processes within the neural network. This materialistic perspective implies a straightforward causal chain: physical processes in the brain lead to conscious experience. However, non-linear temporal models challenge this simplistic view, suggesting a more intricate and potentially reciprocal relationship between consciousness and the temporal flow.

One compelling avenue of inquiry lies in the potential for consciousness to influence the temporal trajectory itself. If the future can influence the past, might conscious intention, belief, or even expectation play a role in shaping the unfolding of events? The phenomenon of self-fulfilling prophecies, mentioned earlier, provides a tantalizing glimpse into this possibility. The very act of believing in a particular future outcome seems to influence present actions, subtly shaping the course of events to align with that belief. This suggests a feedback loop between conscious intention and temporal reality, a dynamic interaction that defies a purely materialistic explanation.

Furthermore, the concept of "retrocausality," where future events influence the past, might be interpreted as an indication of the influence of consciousness on time itself. Imagine a scenario where a future 'decision' – a conscious choice in a future moment – influences the past, thereby altering the conditions that led to that future choice. This raises the possibility that our conscious awareness, even of a future yet-to-be-experienced, may actively participate in creating the very timeline we inhabit. This is not to say that consciousness creates time ex nihilo, but rather that it may have a significant shaping influence on how time unfolds within a non-linear framework.

This perspective aligns with certain interpretations of quantum mechanics, where the observer effect plays a crucial role. The act of observation inevitably alters the observed system, blurring the lines between observer and observed. This is not merely a matter of experimental interference; it suggests a fundamental entanglement between the observer and the reality being observed. This entanglement may extend to the temporal domain, suggesting a dynamic interplay between consciousness and the flow of time. The conscious observer, in this model, is not merely a passive recipient of temporal events, but an active participant in shaping the temporal fabric.

The mind-body problem, a longstanding philosophical debate, also finds new dimensions within the context of non-linear time. The classical dualistic view posits a fundamental separation between mind and body, suggesting that consciousness is something distinct from physical processes. Materialistic views, on the other hand, insist that consciousness is an emergent property of the brain, a complex by-product of neural activity. Non-linear time complicates this dichotomy. If consciousness can influence the physical world by retroactively shaping the past, or by influencing events yet to occur, then the conventional separation between mind and body becomes less distinct.

Instead of a strict causal chain from brain to consciousness, we might envision a more holistic, interconnected system where consciousness and physical processes interact dynamically within the non-linear temporal framework. This does not necessarily imply a supernatural influence, but it suggests that consciousness may have a more active role in shaping reality than traditional materialistic perspectives allow. It calls for a new paradigm, a new model that acknowledges the reciprocal relationship between subjective experience and the temporal dimension. This suggests that the investigation of consciousness requires moving beyond strictly reductionist approaches and embracing the possibility of a more holistic and dynamic interplay between mind, body, and time.

The implications for cognitive science are profound. Traditional cognitive models often focus on the processing of information within the brain, following a relatively linear, sequential model. However, non-linear time suggests a need to reconsider these models. Memories, for example, may not be simply stored as static entities in the brain but might instead be dynamically linked to various points in time, interacting with each other in ways that defy strict linear sequencing. This echoes the work in distributed cognition, highlighting the idea that cognition extends beyond the confines of the brain itself, including the body, environment, and even the temporal dimension.

Furthermore, the concept of non-linear time invites a reassessment of our understanding of perception. If the past, present, and future are interconnected, our perception of reality may not be a simple reflection of present sensory inputs but rather a complex integration of past memories, present sensations, and future expectations. This implies a more dynamic and interactive relationship between perception and the temporal flow, where our conscious experience is not solely determined by present stimuli but shaped by a vast network of temporal influences.

The study of parapsychological phenomena, often dismissed in mainstream science, may also find a more coherent framework within the context of non-linear time. Phenomena such as precognition (the ability to foresee future events) and retrocognition (the ability to perceive past events) are difficult to explain within a strictly linear model. However, if time is not unidirectional, but rather allows for interactions across temporal boundaries, these phenomena become considerably less anomalous. This does not necessarily validate all claims of parapsychological phenomena, but it does suggest that such phenomena deserve a fresh, unbiased evaluation within a framework that acknowledges the possibility of non-linear temporal dynamics.

The investigation into the nature of consciousness within a non-linear temporal framework demands an interdisciplinary approach, drawing upon insights from physics, philosophy, neuroscience, cognitive science, and even parapsychology. This exploration requires a willingness to embrace uncertainty, to challenge long-held assumptions, and to explore unconventional perspectives. The journey is challenging, but the potential rewards are vast: a deeper, more nuanced understanding of consciousness, time, and our place within the cosmos. This new paradigm requires acknowledging the reciprocal interaction between consciousness and the flow of time, rather than merely viewing consciousness as a passive observer of a predetermined temporal stream.

In summary, the relationship between non-linear time and consciousness is a fertile ground for exploration. The introduction of non-linear temporal models compels us to reconsider fundamental assumptions about the nature of subjective experience, the mind-body problem, cognitive processes, and even the validity of parapsychological phenomena. It suggests a far more dynamic and interactive relationship between consciousness and the temporal dimension than previously envisioned, requiring a holistic and interdisciplinary approach to truly grasp its profound implications. This understanding is not only crucial for advancing our scientific knowledge but also for deepening our philosophical comprehension of reality and our role within it. The path forward is paved with challenging questions and the need to confront our biases in order to fully explore this fascinating and potentially transformative frontier of human understanding. The ongoing quest to unlock the secrets of non-linear time holds the potential to revolutionize our understanding of consciousness, leading to a richer, more integrated view of ourselves and the universe.

The implications of a non-linear temporal framework extend far beyond the purely scientific and philosophical realms; they strike at the very core of our existence, shaping our understanding of self, purpose, and our place within the universe. If time is not the immutable, unidirectional river we've always assumed, but a more fluid, interconnected tapestry, then the very foundations of our existential beliefs are called into question. Our concept of personal identity, for instance, is intricately woven into the fabric of linear time. We perceive ourselves as evolving beings, progressing through a series of distinct moments, accumulating experiences and shaping our identities along the way. This narrative of selfhood, however, crumbles under the weight of non-linearity.

If past, present, and future are not strictly ordered, but instead interact and influence each other, our sense of a continuous, evolving self becomes problematic. Could our past selves be influenced by our future decisions, subtly shaping the very experiences that constitute our present identity? Could our future selves, in turn, be influenced by choices made in our past, even unconsciously? This suggests a far more complex and interconnected notion of selfhood than the linear model allows. We may not be the sole authors of our own narratives, but rather participants in a much larger, interwoven temporal drama.

This understanding profoundly affects our sense of free will. If our future actions, to some degree, shape our past, the notion of freely chosen actions becomes ambiguous. Determinism, the philosophical position that all events are predetermined, gains a strange new resonance in a non-linear universe. However, this is not necessarily a bleak deterministic picture; instead, it may suggest a far more intricate and subtle form of determinism, one that acknowledges the feedback loops between past, present, and future, allowing for emergent properties and surprising developments. Rather than a rigid predetermined path, our lives may unfold along multiple possibilities, with our conscious choices interacting with these possibilities to shape the final outcome.

The concept of personal responsibility also takes on new dimensions. If our past actions are subtly influenced by our future intentions, does that diminish our responsibility for those actions? This is a complex ethical question, one that requires careful consideration and a reevaluation of our traditional notions of cause and effect. It may suggest that responsibility lies not in a simple, linear sequence of actions and consequences, but in a web of interconnected influences, where our intentions and choices, regardless of their temporal placement, play a crucial role in shaping our reality.

Our values and beliefs, too, are deeply impacted by the shift to a non-linear temporal framework. If the future can influence the past, then the pursuit of long-term goals takes on a new significance. Our present actions are not merely preparing for a future that is fixed and inevitable, but are actively shaping that future. This underscores the importance of mindful intention and conscious choices, for every action ripples through the temporal fabric, influencing both past and future.

The concept of legacy, the enduring impact we have on the world after we are gone, gains a new layer of complexity. If time is not linear, then our influence extends beyond our physical lifespan. Our actions, thoughts, and intentions may reverberate through time, affecting future generations in ways we cannot fully comprehend. This highlights the immense responsibility we bear, not just for our immediate actions, but for the lasting ripples we create in the temporal tapestry.

Furthermore, the concept of death itself is radically altered within a non-linear temporal framework. If the boundaries between past, present, and future are permeable, the experience of death may not be the absolute end but rather a transition, a shift in temporal perspective. The notion of an afterlife, long debated in philosophical and religious traditions, gains new plausibility within this framework. This is not necessarily a confirmation of any particular religious belief, but a recognition that our current understanding of death is deeply intertwined with our linear perception of time.

Moreover, our understanding of spirituality undergoes a profound transformation. Many spiritual traditions have long emphasized the interconnectedness of all things. A non-linear temporal model powerfully reinforces this concept, suggesting that our experiences, beliefs, and actions are deeply interwoven, not just spatially, but temporally as well. This interconnectivity underscores the profound responsibility we bear towards each other and towards the environment. Our actions are not isolated events, but ripples in a vast, interconnected temporal ocean.

The implications for our social structures and institutions are equally profound. Our legal systems, for instance, are fundamentally based on a linear understanding of causality. The concept of punishment, for example, is rooted in the idea that a past crime demands a present consequence. A non-linear model complicates this, raising questions about the efficacy and justice of punishment if the past and future are dynamically linked. This suggests a need to reconsider our approaches to justice and social order, adopting a more nuanced and holistic perspective that acknowledges the complexities of temporal interconnectedness.

Similarly, our economic systems, often predicated on a linear progression of growth and development, may need re-evaluation. Sustainable development, for example, becomes not simply a matter of future planning but a recognition of the interconnectedness of present actions with past and future consequences. This calls for a more responsible and ethical approach to resource management and economic development, acknowledging the long-term temporal impact of our choices.

The shift to a non-linear understanding of time is not merely a scientific or philosophical curiosity; it is a profound existential challenge. It forces us to reconsider our most fundamental beliefs about ourselves, our place in the universe, and the responsibility we bear towards each other and future generations. This new perspective demands a radical shift in how we view our lives, our actions, and our relationship with the world around us. It is a call for a deeper understanding of our interconnectedness, not just within the present moment, but across the entire temporal landscape. The journey of understanding non-linear time is a journey into the very heart of existence, forcing us to confront the limitations of our current models and embrace a far more intricate and awe-inspiring view of reality.

8 Scientific Exploration of Non-Linear Time

The exploration of non-linear time necessitates a deep dive into the cutting edge of quantum physics. While classical physics operates under the unwavering assumption of a linear, unidirectional time flow, the quantum realm reveals a far more nuanced and perplexing picture. Several key areas of quantum research offer intriguing hints towards a non-linear temporal framework, challenging our fundamental assumptions about causality and the nature of reality.

One of the most compelling areas is the study of quantum entanglement. This phenomenon, famously described by Einstein as "spooky action at a distance," demonstrates a correlation between two or more quantum particles, regardless of the spatial distance separating them. When two entangled particles are measured, their properties are instantaneously correlated, even if they are light-years apart. This instantaneous correlation appears to defy the limitations imposed by the speed of light, suggesting a deeper interconnectedness that transcends the conventional understanding of space and time. Some physicists interpret this as evidence for non-local correlations, implying that entangled particles are not truly separate entities but rather aspects of a single, unified system existing outside the confines of linear spacetime. If this interpretation holds true, it could signify that events seemingly separated by vast temporal distances may be intrinsically linked, defying the strict linearity of cause and effect.

Further bolstering the case for non-linear time are the ongoing investigations into quantum superposition. This principle states that a quantum particle can exist in multiple states simultaneously until measured, at which point it collapses into a single definite state. This "collapse" is a mysterious process, and its precise mechanism remains a subject of intense debate. Some interpretations suggest that the act of measurement itself influences the past, subtly shaping the particle's evolution before the measurement occurred. This retrospective influence on the past contradicts the linear model, suggesting a feedback loop between the future act of measurement and the past state of the particle. This feedback loop blurs the clear distinction between cause and effect, suggesting a more intertwined relationship between past, present, and future. Exploring the intricacies of quantum superposition, therefore, provides valuable insights into the potential for non-linear temporal dynamics.

The concept of quantum tunneling further complicates the linear temporal model. This phenomenon allows particles to pass through energy barriers even if they lack the energy to overcome them classically. This seemingly paradoxical behavior suggests that particles can traverse time in a non-classical manner, potentially bypassing the conventional constraints of linear causality. The implications of quantum tunneling for temporal linearity are profound, as it suggests the possibility of events happening "out of sequence" in a manner that cannot be explained through classical physics. Understanding how quantum tunneling operates may provide crucial clues to the nature of non-linear time and the potential for temporal discontinuities.

The study of quantum decoherence also offers valuable insights into the challenges posed to linear time by quantum phenomena. Decoherence refers to the process by which a quantum system loses its coherence and becomes entangled with its environment. This entanglement with the environment leads to the "collapse" of the superposition into a single state, a process crucial for understanding the transition from quantum to classical behavior. However, the precise mechanism by which decoherence occurs remains a topic of active research. Some theories suggest that decoherence itself involves non-linear temporal processes, where the future state of the system subtly influences its past evolution. This, once again, challenges the conventional notion of a strictly linear causal chain.

Beyond these specific phenomena, the broader theoretical frameworks of quantum mechanics raise further questions about linear time. For instance, the many-worlds interpretation of quantum mechanics proposes that every quantum measurement causes the universe to split into multiple branches, each representing a different outcome. While controversial, this interpretation suggests that all possible timelines exist simultaneously, coexisting in a multi-dimensional temporal landscape. This radical notion fundamentally challenges the concept of a single, linear timeline, offering a picture of a far more complex and multifaceted temporal reality.

Furthermore, the Wheeler-Feynman absorber theory, an alternative approach to electromagnetism, posits a symmetrical interaction between emitter and absorber. This suggests a temporal reciprocity, where the future influences the past as much as the past influences the future. While not directly addressing non-linearity in the same way as quantum entanglement or superposition, this theory highlights a fundamental symmetry in time that challenges the unidirectional arrow of time assumed in classical physics. It suggests a more holistic view of time, where past, present, and future are intertwined and mutually influential, rather than sequentially ordered.

The ongoing research in quantum gravity, seeking to reconcile quantum mechanics with general relativity, offers further potential avenues for exploring non-linear time. This field seeks to understand the behaviour of spacetime at the Planck scale, where the effects of quantum gravity become significant. The very fabric of spacetime may exhibit quantum properties, potentially leading to fluctuations and non-linear behavior. This is a complex and challenging research area, but the potential for discoveries relating to non-linearity is immense. Understanding spacetime at the most fundamental level may reveal previously unimaginable insights into the nature of time itself, moving beyond the linear framework that has underpinned physics for centuries.

The experimental verification of non-linear temporal effects remains a significant hurdle. The delicate nature of quantum phenomena, combined with the challenges of designing experiments capable of detecting subtle temporal anomalies, make it a daunting task. However, ongoing advancements in quantum technologies, such as the development of more sophisticated quantum computers and improved precision measurement techniques, may soon make it possible to directly test for non-linear temporal effects. The development of sensitive detectors capable of detecting minute deviations from linear causality would mark a significant breakthrough in our understanding of time.

The exploration of non-linear time is not merely a scientific endeavor; it has profound philosophical and metaphysical implications. It challenges our understanding of causality, free will, determinism, and even the nature of consciousness itself. A non-linear temporal framework could potentially revolutionize our understanding of the universe, prompting us to re-evaluate our most fundamental assumptions about reality. It suggests a universe far more interconnected and dynamic than we currently comprehend, a universe where past, present, and future are not merely sequentially ordered but are dynamically intertwined, creating a complex tapestry of cause and effect. The implications for our understanding of self, consciousness, and our place in the universe are profound, paving the way for a new, richer understanding of existence. As our ability to probe the quantum realm deepens, and as experimental tools become more sophisticated, we can anticipate a future where the mysteries of non-linear time begin to yield to scientific scrutiny, ultimately leading to a profound shift in our world view. The convergence of quantum physics and the exploration of non-linear time promises a journey into uncharted territory, one that could reshape our understanding of reality itself.

The theoretical frameworks hinting at non-linear time, discussed in the preceding section, are compelling, but ultimately, science demands empirical evidence. The experimental investigation of non-linear time presents formidable challenges. The very nature of time, as we understand it classically, makes it difficult to conceive of experiments capable of directly manipulating or measuring temporal non-linearity. However, progress in quantum technologies offers a glimmer of hope in designing experiments that might detect subtle deviations from the linear temporal paradigm.

One promising avenue lies in the refinement of experiments involving quantum entanglement. As previously discussed, the instantaneous correlation between entangled particles, regardless of distance, appears to violate the speed-of-light limit, suggesting a connection that transcends conventional spacetime. To further investigate this, experiments could be designed to increase the spatial separation between entangled particles, while simultaneously developing more precise methods of measuring their correlated properties. The detection of correlations that occur faster than light could offer direct evidence for non-local influences that are suggestive of a non-linear temporal framework. Furthermore, researchers could explore the possibility of manipulating one entangled particle in a way that demonstrably influences the past state of the other. Such an experiment would require exceptional precision and control over the quantum systems involved, but a successful demonstration would constitute a landmark discovery.

Another area ripe for experimental investigation is quantum superposition. Experiments could focus on refining measurements of quantum systems in superposition, searching for subtle indications that the act of measurement retroactively influences the system's prior state. This would necessitate incredibly precise timing and measurement techniques, capable of capturing even minor deviations from expected linear evolution. One could, for example, design an experiment where a quantum system's state is repeatedly measured over time, looking for statistical anomalies indicative of a retroactive influence. Statistical analysis of large datasets of these measurements might reveal patterns incompatible with a strictly linear temporal sequence. The development of sophisticated algorithms designed to identify such subtle anomalies is crucial for this type of experiment. The challenge here lies not just in the precision of the measurements but also in eliminating all possible sources of classical noise that could mimic non-linear effects.

Quantum tunneling presents yet another opportunity for experimental exploration of non-linear time. Experiments could focus on carefully controlled tunneling events, aiming to quantify any apparent deviation from the expected tunneling rate based on classical physics. Any significant discrepancies could be interpreted as evidence for non-classical temporal processes, where the particle's trajectory deviates from a simple linear progression. However, rigorously disentangling quantum tunneling effects from other classical factors that might influence the results requires meticulous experimental design and control.

The role of quantum decoherence in shaping temporal linearity is also worthy of investigation. Researchers could design experiments that aim to carefully control and manipulate the decoherence process, studying its influence on the evolution of quantum systems. By manipulating the environment that the quantum system interacts with, it might be possible to influence the speed and nature of decoherence, potentially revealing subtle temporal effects not observable under uncontrolled conditions. A successful demonstration of such manipulation would represent a major advance in understanding the interplay between quantum processes and the flow of time.

Moving beyond the realm of individual quantum phenomena, research into the many-worlds interpretation of quantum mechanics could inspire innovative experimental designs. While the direct experimental verification of the many-worlds interpretation remains a formidable challenge, the development of sophisticated quantum computers might offer new avenues of exploration. The ability to simulate complex quantum systems could allow researchers to test predictions of the many-worlds interpretation and potentially reveal experimental signatures that could indirectly support or refute this controversial interpretation. Such simulations could focus on the evolution of quantum systems under different measurement scenarios, looking for statistical evidence consistent with the branching timelines predicted by this theory.

Moreover, the experimental exploration of non-linear time requires the development of new measurement tools and technologies. The challenge lies in creating devices capable of detecting incredibly subtle deviations from linear temporal behavior. Advances in quantum metrology—the science of precise measurement—could offer significant contributions. High-precision clocks, for instance, might be able to detect incredibly tiny fluctuations in the rate of time, potential indicators of temporal non-linearity. Similarly, the development of highly sensitive detectors capable of capturing minute variations in the fabric of spacetime might reveal temporal anomalies otherwise undetectable.

The limitations to experimental investigations of non-linear time are significant. Quantum systems are inherently fragile, and controlling them with the required level of precision is extremely difficult. External noise can easily mask or mimic subtle temporal effects, making it crucial to design experiments that minimize interference from the environment. Furthermore, the interpretation of experimental results is often complex and ambiguous, requiring sophisticated statistical analyses and theoretical modeling. It is critical to design experiments that rigorously control for confounding variables, and ensure that any observed effects cannot be attributed to classical phenomena.

Finally, it's crucial to acknowledge the philosophical and interpretational challenges involved in any experimental attempts to verify non-linear time. Our ingrained assumptions about causality and the nature of time can significantly influence the design and interpretation of experiments. It's essential to approach this research with an open mind, acknowledging the possibility that our current conceptual frameworks might need significant revision in light of experimental findings. The potential for unexpected results and radical reinterpretations of our understanding of reality underscores the immense significance of this field of inquiry.

Despite the considerable hurdles, the pursuit of experimental evidence for non-linear time is a compelling scientific endeavor with potentially revolutionary implications. As our understanding of quantum physics deepens and our technological capabilities advance, the possibility of experimentally verifying these hypotheses becomes increasingly realistic. The convergence of quantum technologies, advanced metrology, and sophisticated theoretical frameworks offers a path towards understanding the subtle complexities of time itself, potentially transforming our view of reality and our place within it. The journey is daunting, but the potential rewards – a profound reimagining of our understanding of the universe – are immeasurable.

The theoretical exploration of non-linear time, while fascinating, only scratches the surface of its potential implications. If we were to successfully demonstrate and harness non-linear temporal effects, the technological ramifications would be nothing short of revolutionary, impacting nearly every facet of our lives. Consider the profound implications for computing. Classical computers operate sequentially, processing information step-by-step in a linear fashion. However, if we could exploit non-linear temporal effects, we could potentially design computers that process information across multiple temporal points simultaneously. Imagine a computer capable of performing calculations not just in the present, but also in the past and future, effectively bypassing the limitations of sequential processing. Such a machine could possess computational power exceeding anything we can currently envision, ushering in an era of unprecedented technological advancement.

The development of such a "non-linear computer" presents a formidable challenge. It would necessitate a deep understanding of the mechanisms governing non-linear time, coupled with the ability to manipulate these mechanisms at a quantum level. We might need to engineer devices capable of creating and controlling localized regions of spacetime exhibiting non-linear temporal properties, effectively constructing "temporal circuits" that allow for non-sequential information processing. The design and construction of such circuits would require significant advances in nanotechnology and quantum control, enabling the precise manipulation of quantum systems to achieve the desired temporal effects. The architecture of this new type of computer would fundamentally differ from today's designs, potentially relying on entanglement or quantum superposition to perform computations across multiple temporal points. The programming paradigms would also need a radical overhaul, requiring new languages and algorithms capable of harnessing the power of non-linear temporal computation.

Beyond computation, the implications for communication are equally profound. If information could be transmitted across different temporal points, we could potentially achieve instantaneous communication across vast distances, effectively eliminating the limitations imposed by the speed of light. This could revolutionize global communication, allowing for real-time interactions across the planet and beyond, with no transmission delays. Imagine instantaneous global collaboration, teleconferencing without latency, and near-instantaneous data transmission across interstellar distances. This technology, however, would require the ability to manipulate temporal connections and entangle information across different temporal points, requiring a level of control over quantum systems currently beyond our capabilities.

The energy sector would also undergo a radical transformation. If we could access and manipulate energy from different temporal points, we could potentially create devices that extract energy from the past or future, bypassing the constraints of present-day energy sources. This could lead to virtually limitless energy production, resolving the global energy crisis once and for all. However, the theoretical and practical challenges involved in developing such technology are immense, requiring a thorough understanding of the thermodynamics of non-linear time and the development of devices capable of manipulating energy flow across different temporal points. Careful consideration of potential paradoxes and unintended consequences would be crucial, ensuring the responsible and sustainable utilization of such powerful technology.

Moreover, the implications extend far beyond these specific sectors. Medical applications could benefit immensely from the ability to manipulate temporal processes. For instance, we could potentially develop new diagnostic and therapeutic techniques based on the ability to access and influence temporal events at the molecular or cellular level. Imagine non-invasive diagnostic methods that could detect disease before symptoms manifest, or therapeutic interventions that could reverse the progression of age-related diseases or even injuries at a cellular level. The ethical considerations of such technology would, however, be paramount. The ability to influence the past could raise serious ethical questions concerning free will and causality.

The implications for transportation are equally transformative. If we could manipulate spacetime to reduce travel times, or even allow for travel across different temporal points, we could revolutionize travel, enabling near-instantaneous transportation across vast distances. This would have a profound impact on global connectivity, fostering unprecedented levels of cross-cultural exchange and collaboration. Again, the ethical considerations of such technology are profound, requiring careful consideration of potential paradoxes and unforeseen consequences.

However, alongside the boundless potential, the technological manipulation of non-linear time presents considerable challenges and risks. The potential for unintended consequences, including the creation of temporal paradoxes, requires careful consideration. The development of such technology must be guided by rigorous ethical frameworks, prioritizing safety and responsible innovation. The potential for misuse is significant, requiring strict regulatory oversight to prevent catastrophic misuse. The development of such technologies must involve extensive interdisciplinary collaboration, bringing together physicists, engineers, ethicists, and policymakers to ensure responsible innovation.

The exploration of non-linear time is not just a scientific pursuit, it's a philosophical and ethical one. The very foundations of our understanding of causality and free will are challenged by the prospect of manipulating time. Understanding the implications of non-linear time requires grappling with profound philosophical questions about the nature of reality and our place within it. The development of technologies based on non-linear time calls for a nuanced approach that integrates scientific rigor, ethical considerations, and societal impact assessments. It requires not only scientific expertise, but a commitment to responsible technological advancement, ensuring that these potentially transformative tools are used for the benefit of humanity. The journey into this new frontier must be marked by caution, foresight, and a commitment to thoughtful innovation.

The path ahead is fraught with difficulties, but the potential rewards are immense. The exploration and potential harnessing of non-linear temporal phenomena could fundamentally reshape our technological landscape, profoundly impacting every aspect of our lives. From fundamentally altering computation and communication to revolutionizing energy production and medical treatment, the implications are vast and far-reaching. However, the successful navigation of this uncharted territory requires not just scientific breakthroughs, but a deep and comprehensive understanding of the ethical, philosophical, and societal consequences of manipulating the very fabric of time itself. This is a journey that requires collaboration across disciplines, thoughtful consideration, and a commitment to responsible innovation. Only then can we hope to unlock the immense potential of non-linear time without succumbing to its inherent risks.

The prospect of manipulating non-linear time, as discussed previously, presents not just technological hurdles but also epistemological and ethical challenges of unprecedented magnitude. Successfully navigating this uncharted territory requires a concerted effort transcending the boundaries of traditional scientific disciplines. The exploration of non-linear time demands a collaborative approach, integrating the perspectives and methodologies of physicists, philosophers, psychologists, and potentially even specialists in fields like parapsychology and cognitive science. This interdisciplinary approach is not merely desirable; it is absolutely essential for responsible and effective advancement in this field.

Physicists, of course, play a crucial role in providing the foundational understanding of the physical mechanisms governing time. Their expertise in quantum mechanics, relativity, and cosmology is vital in formulating testable hypotheses and developing experimental methodologies to investigate non-linear temporal phenomena. However, the implications of non-linear time extend far beyond the realm of physics. The very concepts of causality and free will, cornerstones of philosophical inquiry for centuries, are fundamentally challenged by the possibility of manipulating temporal sequences. Philosophers can contribute significantly by analyzing the logical and ethical implications of altering the past or influencing the future. What are the ramifications for personal identity if one could alter past choices? What are the ethical implications of creating alternate timelines or interfering with the causal chain of events? These are questions demanding rigorous philosophical scrutiny.

Furthermore, the human experience of time is a complex and multifaceted phenomenon, shaped by psychological factors and cognitive processes. Psychologists and cognitive scientists can offer invaluable insights into how our perception and understanding of time are constructed and how they might be affected by the manipulation of non-linear temporal phenomena. Studies exploring our subjective experience of time, our ability to mentally time travel (e.g., through reminiscence or future planning), and the potential for altered states of consciousness to influence our perception of time are all relevant in understanding the implications of any technological intervention. Their expertise can inform the design of experiments to measure the psychological and cognitive effects of exposure to non-linear temporal fields, helping us assess the potential impact on human consciousness and behavior.

The potential involvement of fields like parapsychology, while controversial in mainstream scientific circles, merits consideration. Parapsychology explores phenomena, such as precognition and retrocausation, that seem to imply some form of interaction with non-linear time. While the evidence base in parapsychology remains highly debated, the possibility that these seemingly anomalous phenomena might be explained by underlying non-linear temporal mechanisms cannot be entirely dismissed. A careful, critical analysis of parapsychological research, employing rigorous scientific methodology, could potentially illuminate aspects of non-linear time that remain hidden within conventional scientific frameworks. This does not imply a wholesale acceptance of all parapsychological claims, but rather a reasoned exploration of their potential relevance to a broader understanding of non-linear time.

Beyond the core disciplines, other fields can contribute significantly. Engineers are essential for designing and building the necessary instruments and technologies to detect and manipulate non-linear temporal phenomena. Their expertise in nanotechnology, quantum control, and advanced materials science is crucial for creating devices capable of generating and manipulating localized regions of spacetime with non-linear temporal properties. Computer scientists would be instrumental in developing new computational models and algorithms capable of harnessing the power of non-linear temporal computation, creating the programming languages necessary to interact with such advanced systems. The legal and ethical implications also necessitate collaboration with ethicists and policymakers to establish responsible guidelines for research and development. This includes developing ethical frameworks to govern the use of any technologies that emerge, mitigating potential risks and ensuring that these powerful tools are used for the benefit of humanity.

The collaborative process itself is crucial. It requires the establishment of interdisciplinary research teams, collaborative research platforms, and regular cross-disciplinary dialogues. This means not only bringing together experts from different fields, but also creating an environment where open communication and mutual respect are fostered. This open dialogue can be facilitated through workshops, conferences, and the publication of interdisciplinary research. A crucial element of this collaboration is the creation of shared conceptual frameworks and research methodologies that integrate the perspectives of multiple disciplines. This might involve developing new theoretical models capable of encompassing both physical and psychological aspects of time, or establishing shared standards for conducting experiments that test the validity of hypotheses drawn from different disciplinary perspectives.

Moreover, the potential societal impact of manipulating non-linear time is profound and requires careful consideration. The development of technologies based on non-linear time could lead to transformative changes in nearly every aspect of human life, impacting our social structures, economic systems, and cultural values. Sociologists, anthropologists, and other social scientists can provide insights into the societal impacts of such technologies and help in anticipating and mitigating potential risks. Furthermore, it is vital to consider the distribution of access to these technologies and to ensure that the benefits are shared broadly across society. This requires careful consideration of equity and social justice implications, ensuring the equitable distribution of resources and minimizing the risk of exacerbating existing inequalities.

The challenges are immense, but the potential rewards justify this ambitious interdisciplinary approach. The successful exploration and harnessing of non-linear temporal phenomena hold the promise of resolving some of humanity's most pressing challenges and advancing our understanding of the universe and our place within it. It is a journey into uncharted territory that requires not just scientific breakthroughs, but a fundamental shift in our approach to scientific inquiry, embracing collaboration, careful consideration, and a commitment to responsible innovation. Only through such a concerted, multi-faceted approach can we navigate the complexities of non-linear time and unlock its vast potential while mitigating its inherent risks. The success of this endeavor depends upon the synergy created by the diverse perspectives and expertise brought to bear by this cross-disciplinary collaboration, emphasizing not only scientific rigor, but also ethical reflection, societal responsibility, and a profound understanding of the human condition in the context of time itself. The future of this field hinges on this collaborative, multifaceted approach. It's a future ripe with potential, but one that must be approached with both ambition and caution.

The exploration of non-linear time is still in its nascent stages, akin to the early days of quantum mechanics or relativity. While theoretical frameworks are beginning to emerge, the experimental verification and practical application remain largely in the realm of speculation. This necessitates a focused and multi-pronged approach to future research, one that transcends disciplinary boundaries and fosters synergistic collaboration.

One crucial area is the development of more sophisticated theoretical models. Current models, often built upon interpretations of quantum mechanics and general relativity, require further refinement and integration.

We need models capable of not just describing non-linear temporal phenomena but also predicting them with a high degree of accuracy. This involves exploring the potential connections between quantum entanglement, spacetime curvature, and the subjective experience of time. For instance, could entangled particles exhibit correlations that transcend the conventional arrow of time, offering a tangible manifestation of non-linear temporal effects? Could the curvature of spacetime, as predicted by general relativity, be manipulated to create localized regions with altered temporal properties? These are fundamental questions that require further investigation.

Furthermore, the development of novel experimental methodologies is paramount. Current technological limitations hinder the direct observation and manipulation of non-linear temporal effects. Advances in quantum technologies, such as quantum computing and quantum sensing, hold immense promise in this regard. Quantum computers, with their ability to simulate complex quantum systems, could be utilized to model and test theoretical predictions about non-linear time. Quantum sensors, possessing unprecedented sensitivity, could potentially detect subtle temporal anomalies that remain invisible to conventional instruments. The development of novel materials with unique properties, such as metamaterials capable of manipulating electromagnetic fields in unconventional ways, could also play a critical role in designing experiments to probe non-linear temporal phenomena. Furthermore, advancements in precision measurement techniques, coupled with rigorous statistical analysis, are essential for ensuring the reliability and replicability of experimental results.

Beyond purely physical approaches, investigating the subjective experience of time through rigorous psychological and cognitive science experiments is crucial. This involves refining methodologies for measuring subjective time perception, exploring the relationship between altered states of consciousness (e.g., meditation, hypnagogia) and the experience of time, and investigating the potential for neurofeedback or other brain stimulation techniques to influence our perception of temporal flow. Are there specific brain regions or neural pathways primarily responsible for our linear perception of time? If so, could manipulating these areas using neuroscientific techniques alter our experience of time, potentially revealing underlying non-linear temporal dynamics? These questions require innovative research paradigms involving neuroimaging, psychophysical experiments, and advanced data analytic techniques.

The intersection of parapsychology and the investigation of non-linear time remains a contentious yet potentially fruitful area of inquiry. While the claims of parapsychology often lack robust empirical support, dismissing them entirely would be premature. A rigorous, skeptical examination of parapsychological phenomena, such as precognition and retrocausation, using sophisticated experimental designs and statistical analysis, may reveal subtle correlations that provide clues about underlying non-linear temporal mechanisms. This necessitates the development of new experimental protocols capable of minimizing biases and controlling for confounding variables, ensuring that any purported evidence of non-linear temporal effects meets the highest standards of scientific rigor. The goal here is not to validate all parapsychological claims but rather to critically examine whether any of these phenomena offer empirical support for theoretical models of non-linear time.

The ethical implications of manipulating non-linear time must be addressed proactively, not reactively. The possibility of altering the past, influencing the future, or creating alternate timelines raises profound ethical concerns that demand careful consideration by ethicists, policymakers, and the wider scientific community. We must establish robust ethical guidelines and regulatory frameworks to govern research in this field, ensuring that the pursuit of knowledge does not come at the cost of human well-being or societal stability. This involves not only defining ethical boundaries for experimentation but also addressing the potential societal impact of technologies based on non-linear time. What are the implications for personal identity if one could alter one's past? How do we ensure the equitable distribution of any technologies that emerge, preventing the exacerbation of existing social inequalities? These are complex societal questions demanding a multidisciplinary approach involving ethicists, sociologists, legal scholars, and policymakers.

Furthermore, the development of robust computational models is essential for harnessing the potential of non-linear temporal phenomena.

This requires breakthroughs in computer science, particularly in the development of new algorithms and programming languages capable of processing information in non-linear temporal frameworks. The creation of temporal databases, capable of storing and retrieving information across multiple timelines or temporal dimensions, represents a significant computational challenge that needs to be tackled. Similarly, the development of sophisticated simulation tools capable of predicting the consequences of altering temporal sequences, both at the individual and societal levels, is essential for responsible innovation.

The collaborative nature of this research is paramount. The establishment of interdisciplinary research centers, fostering collaborations between physicists, philosophers, psychologists, computer scientists, ethicists, and policymakers, is a crucial step. This necessitates a shift in the current scientific paradigm, moving away from disciplinary silos towards a more integrated and synergistic approach. Regular interdisciplinary workshops, conferences, and collaborative research projects can facilitate the exchange of ideas and knowledge, fostering a shared understanding of the challenges and opportunities presented by non-linear time.

The long-term vision should involve developing a unified theory of time, one that integrates the physical, psychological, and philosophical perspectives. This unified theory would not only provide a coherent explanation for non-linear temporal phenomena but also guide future research directions and technological development. The implications of such a theory are far-reaching, with the potential to revolutionize our understanding of the universe, the nature of consciousness, and our place within the cosmos.

Finally, it is essential to recognize the inherent uncertainties involved in this field. The exploration of non-linear time is a journey into uncharted territory. The possibility of encountering unexpected and potentially unpredictable consequences is a reality that needs to be acknowledged and carefully considered. This necessitates a cautious, responsible, and ethically informed approach to research and development, ensuring that the potential benefits of manipulating non-linear time are carefully weighed against the potential risks. The future of this field depends not only on scientific breakthroughs but also on ethical reflection, responsible innovation, and a profound understanding of the complex interplay between science, society, and the very nature of time itself.

9 The Physics of Time Perception

Our experience of time, that seemingly continuous flow from past to present to future, is not a direct reflection of a physical reality but rather a construct of our brain. While physicists grapple with the nature of time itself – its linearity, its potential non-linearity, and its relationship to space – neuroscientists are meticulously mapping the intricate neural pathways that shape our subjective perception of its passage. This subjective experience, far from being a simple reflection of objective clock time, is a dynamic and malleable phenomenon influenced by a multitude of factors, including attention, emotion, and even our expectations.

The brain doesn't possess a single "time-keeping" center. Instead, a complex interplay of multiple brain regions contributes to our temporal awareness. The cerebellum, often associated with motor control and coordination, plays a significant role in our perception of short time intervals, those lasting mere seconds or fractions of a second. Studies using transcranial magnetic stimulation (TMS), a non-invasive brain stimulation technique, have demonstrated that disrupting cerebellar activity can significantly alter the perception of brief temporal durations. This suggests that the cerebellum acts as a kind of internal clock, constantly monitoring and integrating sensory information to establish a sense of temporal order in our immediate environment.

However, our perception of longer time intervals is a different story altogether. The prefrontal cortex, the brain's executive control center, is heavily involved in our understanding of extended time spans. This region is responsible for higher-level cognitive functions such as planning, decision-making, and working memory, all of which are inextricably linked to our perception of time's passage. Damage to the prefrontal cortex can lead to significant impairments in time estimation, demonstrating its crucial role in our long-term temporal awareness.

The hippocampus, a structure deeply implicated in memory formation and spatial navigation, also plays a critical role in time perception. The hippocampus helps to contextualize events within a temporal framework, connecting memories and experiences in a chronological sequence. This process of memory consolidation is crucial for our sense of a continuous, unfolding narrative of our lives. Lesions to the hippocampus often result in difficulties in remembering the temporal order of events, highlighting its importance in our construction of personal timelines.

Beyond these core regions, numerous other brain structures contribute to our perception of time. The striatum, part of the basal ganglia, influences time perception by integrating reward and motivation signals. The amygdala, the emotional processing center of the brain, can dramatically distort our perception of time, especially during moments of intense fear or excitement. In such situations, time can seem to slow down or speed up, a phenomenon likely driven by the amygdala's influence on the release of stress hormones like cortisol. These hormones can alter the activity of other brain regions involved in time processing, effectively creating a subjective distortion of the temporal flow.

The neurochemical underpinnings of time perception are equally complex. Dopamine, a neurotransmitter associated with reward and motivation, plays a crucial role in our ability to accurately estimate time intervals. Studies have shown that manipulating dopamine levels can significantly impact time perception, suggesting that this neurotransmitter plays a pivotal role in the brain's internal clock mechanisms. Other neurotransmitters, such as serotonin and norepinephrine, also appear to be involved, although their exact roles are still under investigation.

Our understanding of the neurological basis of time perception is constantly evolving. Advances in neuroimaging techniques, such as fMRI and EEG, have enabled researchers to obtain increasingly detailed maps of brain activity during time perception tasks. These studies have provided valuable insights into the neural networks involved and the complex interactions between different brain regions. However, many questions remain unanswered. The precise mechanisms by which the brain converts sensory inputs into a subjective experience of time remain elusive. The interplay between different brain regions and neurochemical pathways involved is complex and requires further investigation.

Furthermore, the interaction between objective and subjective time presents a significant challenge. While our internal clocks attempt to measure the passage of objective time, our perception of this passage is readily influenced by subjective factors. Attentional demands, for instance, can significantly impact our perception of time. When we're deeply engaged in a task, time often seems to fly by, whereas moments of boredom can stretch seemingly endlessly. This highlights the active and dynamic nature of our brain's construction of time.

Beyond the effects of attention, emotion plays a powerful role in shaping our subjective experience of time. Moments of intense emotion, whether positive or negative, often distort our perception of time's passage. A traumatic event may feel like it lasted an eternity, while a joyful experience can feel fleeting. This suggests that the brain's emotional processing systems interact dynamically with its temporal processing mechanisms to shape our subjective experience of duration.

The potential for external manipulations of time perception is a fascinating area of ongoing research. Techniques like transcranial direct current stimulation (tDCS), another non-invasive brain stimulation method, have shown promise in modulating time perception. Studies using tDCS have shown that stimulating specific brain regions can either shorten or lengthen the perception of time, offering tantalizing glimpses into the possibility of actively controlling our subjective experience of temporal flow.

However, the ethical implications of such interventions require careful consideration. Manipulating our perception of time could have significant effects on our decision-making, our memory, and even our sense of self. The potential for misuse of these technologies raises concerns that necessitate careful ethical review and regulatory oversight.

The study of time perception at the neurological level is not merely an academic exercise. Understanding the brain mechanisms that underpin our experience of time has profound implications for various aspects of human life. Improving our understanding of these mechanisms can assist in treating conditions like attention-deficit hyperactivity disorder (ADHD), where time perception is often impaired. It can also lead to advancements in therapies for individuals suffering from anxiety disorders, where the experience of time is frequently distorted.

Moreover, unraveling the neural mechanisms of time perception could have implications for our understanding of consciousness. Our subjective experience of the flow of time is an integral aspect of our conscious awareness. By better understanding how our brain generates this experience, we might gain valuable insights into the nature of consciousness itself.

Ultimately, the neurological basis of time perception is a frontier of research rich with both scientific and philosophical implications. The integration of findings from neuroscience, psychology, and physics is essential for building a comprehensive understanding of this complex phenomenon. This integrated approach will not only enhance our understanding of the intricate workings of the human brain but also provide insights into the fundamental nature of time itself and its role in shaping our reality. The journey toward a complete understanding remains long and challenging, but the potential rewards—both scientifically and philosophically—are immense. The more we learn about how the brain constructs our perception of time, the closer we come to understanding the very fabric of our experience, and perhaps, the universe itself.

Our exploration of time perception thus far has focused on the neurological mechanisms underlying this fundamental aspect of human experience. However, the brain's construction of time is not solely determined by its intricate wiring and neurochemical cascades. Psychological factors, deeply intertwined with our cognitive processes and emotional states, exert a profound influence on how we experience the passage of time. This subjective experience, often dramatically different from the objective measurement of time provided by clocks, is a complex interplay of neural activity and psychological processes.

One of the most significant psychological influences on time perception is our attentional state. When fully engrossed in an engaging activity, time often seems to fly by. This phenomenon, commonly referred to as "time dilation," is not a distortion of objective time, but rather a reflection of how our focused attention impacts our internal temporal processing. When our attention is sharply focused, our brain processes information more efficiently, creating the illusion that less time has elapsed than is objectively the case. Conversely, during periods of boredom or inactivity, time can seem to stretch out endlessly, each moment dragging on as our minds wander and search for stimulation. This subjective stretching of time stems from the brain's relative under-engagement, leading to a heightened awareness of the passage of each moment.

Consider the experience of waiting for an anticipated event. The anticipation itself can profoundly influence our perception of time. The longer the wait, the more intensely we focus on the passage of time, exacerbating the feeling that time is moving slowly. This highlights the close relationship between expectation and time perception; the stronger the anticipation, the slower time appears to move. Conversely, if we are unexpectedly occupied during a wait, the period of waiting can feel significantly shorter because our attention is redirected, minimizing our awareness of the passage of time.

The role of memory in shaping our perception of time is equally significant. Our memories of past events are not simply snapshots of moments in time, but rather reconstructions colored by our emotional and cognitive states at the time of the experience. Consequently, vivid and emotionally charged memories often feel like they occurred more recently than they actually did, creating a temporal distortion where the past seems closer to the present. Conversely, memories that lack emotional salience or are poorly encoded tend to recede further into the past, further distorting our perception of the temporal distance between events.

This memory-based temporal distortion is particularly relevant in the context of autobiographical memory. Our recollections of life experiences are rarely linear or accurate accounts of objective time. Instead, our personal timelines are constructed by selectively retrieving and reconstructing memories, often influenced by current emotions and cognitive biases. This process is not necessarily flawed; it serves a crucial function in maintaining a coherent and meaningful sense of self across time. However, it invariably leads to a subjective and often inaccurate perception of the temporal sequence and duration of events in our lives.

Emotional state exerts a substantial influence on our perception of time. Moments of intense emotion, be it joy, fear, or sorrow, often lead to a dramatic distortion of time perception. A near-miss accident, for instance, can feel like it stretched out over an extended period, while a fleeting moment of intense joy can feel compressed, leaving an impression of having lasted only an instant. This emotional influence on time perception is well-documented and reflects the dynamic interplay between the brain's emotional processing centers, such as the amygdala, and its temporal processing networks. The heightened activity in the amygdala during emotional experiences may alter the rate at which the brain processes information, resulting in the subjective perception of time dilation or compression.

The subjective experience of time is also influenced by the pace of life and the density of events. A day packed with numerous and varied activities can subjectively feel shorter than a day spent in inactivity, even if the objective duration of time remains identical. This reflects the brain's tendency to evaluate temporal duration based not only on the clock time elapsed, but also on the richness and diversity of experiences encoded in memory. A monotonous routine can, therefore, create an exaggerated sense of time's duration, while a vibrant and dynamic lifestyle may compress our experience of time's passage.

Several experimental paradigms have been developed to explore subjective time perception. These studies typically involve presenting participants with temporal stimuli, such as brief sounds or visual cues, and asking them to estimate their duration. Researchers have employed a range of techniques, from simple reproduction tasks (where participants are asked to reproduce a given duration) to more complex estimation tasks (where participants are asked to judge the relative duration of different stimuli). These experiments have consistently demonstrated that subjective time perception is not simply a passive recording of objective time, but rather an active and malleable process influenced by various psychological factors.

Studies involving time reproduction tasks, for example, have revealed systematic biases in time estimation. Overestimation of short durations and underestimation of long durations are common findings, highlighting the non-linear relationship between subjective and objective time perception. This non-linearity is not simply an artifact of experimental design, but likely reflects the limitations of the brain's internal clock mechanisms and the influences of various cognitive and emotional factors.

In addition to experimental studies, anecdotal evidence and personal experiences vividly illustrate the subjectivity of time perception. The common observation of time seeming to "fly by" during enjoyable activities or "drag on" during monotonous tasks speaks to the powerful influence of our emotional and attentional states on the internal experience of time.

Furthermore, experiences under conditions of high stress or intense emotion further highlight this malleability, with traumatic events often perceived as lasting far longer than they objectively did. This subjective stretching of time serves as a powerful coping mechanism, allowing the individual to process and integrate a traumatic experience more effectively.

The subjective nature of time perception has implications beyond our daily experiences. It can affect how we make decisions, how we plan for the future, and how we evaluate past events. In legal contexts, for instance, witness testimony regarding the duration of events is known to be highly subjective and unreliable, with the accuracy of temporal estimations being significantly influenced by emotional factors and the overall stress level during the events in question.

Furthermore, understanding the psychological factors affecting time perception can offer valuable insights into various mental health conditions. Conditions like ADHD (Attention Deficit Hyperactivity Disorder) are frequently associated with difficulties in time estimation and management, suggesting that impairments in attentional processes may underlie this temporal deficit. Similarly, anxiety disorders can involve distorted perceptions of time, with individuals experiencing heightened anxiety often perceiving time as moving more slowly. Investigating these psychological influences offers potential avenues for the development of more effective therapeutic interventions.

In conclusion, our perception of time is not a direct reflection of physical reality but a complex subjective experience constructed by our brain. The psychological factors discussed here, namely attention, memory, and emotion, actively participate in this construction, shaping how we experience the seemingly continuous flow of time. Understanding this interplay between neurological and psychological processes is vital not only for a complete understanding of time perception itself but also for applications across various fields, from legal proceedings to mental health treatment. The continuous study and refinement of our understanding of these intricate mechanisms will undoubtedly yield further insights into the fundamental nature of human experience and the subjective reality we construct moment by moment.

Our previous discussion highlighted the profound influence of attention, memory, and emotion on our subjective experience of time. However, the complexities of time perception extend further, encompassing a range of cognitive biases that systematically distort our estimations of temporal duration and sequencing. These biases, often operating unconsciously, reveal a fascinating interplay between our internal clock and the sophisticated, yet fallible, cognitive mechanisms that shape our perception of reality.

One prominent cognitive bias affecting time perception is *duration neglect*. This bias refers to the observation that our overall evaluation of an experience is less influenced by its total duration than by its most intense moments. Consider two experiences: a ten-minute massage that was consistently pleasant, and a twenty-minute massage that started pleasantly but ended with an uncomfortable pressure point. While the second massage lasted twice as long, many people would rate the first massage as more enjoyable. The reason lies in duration neglect: the overall judgment prioritizes the peak intensity of pleasure, overshadowing the contribution of the total duration. This demonstrates that our memory of an experience, rather than its actual temporal span, significantly determines our retrospective evaluation. This has implications beyond spa treatments; it influences how we recall vacations, conversations, and even significant life events. The more intense a moment was, regardless of its length, the more heavily it weighs in shaping our memory of the overall experience.

Closely related to duration neglect is the *peak-end rule*. This heuristic suggests that our retrospective evaluation of a painful or unpleasant experience is determined primarily by the peak intensity of the unpleasantness and the final intensity at the end of the experience. The average level of pain throughout the duration is relatively unimportant. Studies involving colonoscopies have vividly illustrated this phenomenon. Patients who underwent a longer procedure but experienced a gradual reduction in discomfort at the end rated their overall experience as significantly less unpleasant than patients who underwent a shorter procedure with a sharply painful ending. This finding has profound implications for the design of medical procedures and other experiences intended to minimize negative impact. Understanding the peak-end rule allows for the strategic modulation of the intensity of unpleasant experiences to optimize retrospective evaluation, even if the overall duration cannot be altered. The application of this principle extends far beyond medical scenarios. It helps in understanding consumer experiences, work-related stress, and even relational dynamics where the culmination of a conversation or interaction can greatly impact its overall perception.

The influence of cognitive biases on our perception of linearity is also noteworthy. Our mental representation of time rarely adheres to a strictly linear model. Memory often presents events not in a chronological sequence but rather as a network of interconnected episodes, triggered by associative links and emotional connections. This nonlinear representation can lead to distortions in our temporal judgments, creating anomalies in our sense of how events unfolded. Imagine recounting a childhood memory; the narrative might not flow seamlessly from one event to the next but might instead jump across years, connecting seemingly disparate moments through shared themes or emotional resonance. This non-linearity is not simply a result of faulty memory retrieval, but rather reflects a cognitive strategy designed to make sense of a vast and complex body of experiences. The brain does not store memories as a chronologically ordered list, but rather as a multifaceted network, and this network topology heavily influences our time perception.

Furthermore, the contextual framing of an event can dramatically impact our perception of its duration. Consider the difference between a five-minute wait in a silent, empty waiting room versus a five-minute wait while engrossed in a captivating book. The latter may feel far shorter. This is because our cognitive resources are consumed by the engaging activity, reducing our awareness of the passage of time. The context, therefore, influences our perception by modulating our attentional resources and hence, the brain's ability to accurately track time's passage. This implies that subjective time perception is not solely a function of an internal clock but is tightly coupled with our cognitive processing of the environment and the allocation of our attentional resources. The less attention we dedicate to monitoring the passage of time, the more distorted our perception becomes. This context-dependent temporal distortion is not limited to simple waiting periods; it affects our perception of longer events, as well. A busy workday, for instance, can feel much shorter than an equally long period spent in tedious inactivity despite the objectively equal duration.

The influence of expectations further complicates the relationship between our subjective experience and objective measurement of time. Anticipation can dramatically distort our perception of waiting periods. A highly anticipated event can make the wait seem interminable, while an unexpected delay might go largely unnoticed. This is because our anticipation is an active cognitive process, actively focusing our attention on the time remaining, thereby amplifying our awareness of its passage. The strength of anticipation directly correlates with the perceived duration of the wait. This also reveals a complex interaction between our internal clock and our emotion regulation systems; anxiety related to anticipation can directly distort our perception of time's flow.

Moreover, the perceived rate of time flow itself can be subject to cognitive biases. Our perception of time is often described as faster during childhood and slower during old age. This observation, however, is partly a function of how we measure and remember time, and partly a result of the density of new experiences encountered at different life stages. Childhood, brimming with novel experiences and learning, provides a richer tapestry of memories, making the perceived duration seem compressed in retrospect. In contrast, as we age, the novelty of experiences decreases, and events may seem to blend together, resulting in the perception of a slower rate of time. It's crucial to note that this isn't a change in the objective flow of time, but a shift in how our brain processes and encodes the temporal richness of experiences across the lifespan.

These cognitive biases, along with the previously discussed factors of attention, memory, and emotion, collectively paint a picture of time perception that is far from a passive reflection of objective reality. Our experience of time is an active construction, influenced by a complex interplay of neural mechanisms, cognitive processes, and affective states. The linearity of time, often assumed in our daily lives, is not so neatly upheld by the reality of human perception; our mental timelines are malleable, distorted, and context-dependent. This profound subjectivity has vast ramifications, impacting our personal narratives, our decisions, our legal systems, and our understanding of our own consciousness. Future research into the neural and cognitive underpinnings of these biases promises deeper insights into the elusive nature of time itself and the intricately constructed nature of subjective experience.

The subjective experience of time's passage is profoundly shaped by the nature of memory itself. While we often assume a linear, chronological unfolding of events, our memories rarely adhere to such a rigid structure. Memory is not a passive recording device; it is an active reconstructive process, constantly shaping, reshaping, and reinterpreting our past experiences. This active reconstruction inherently introduces the possibility of distortion, affecting our perception of temporal duration and sequencing.

Consider the phenomenon of telescoping, where memories of recent events are perceived as further in the past than they actually are, while distant memories seem closer than they truly are. This distortion is not simply a matter of faulty recall; it reflects the dynamic nature of memory consolidation and retrieval. Recent events are still being integrated into our long-term memory networks, making their temporal placement less certain. Conversely, frequently recalled or emotionally salient memories, regardless of their actual age, maintain a heightened accessibility, thereby seeming more recent. This telescoping effect demonstrates that our perception of temporal distance is not a direct reflection of the objective passage of time but rather a function of memory's organizational processes.

The role of emotion in shaping temporal perception through memory is equally significant. Highly emotional events, whether positive or negative, tend to be encoded more vividly and retrieved more readily than neutral experiences. The amygdala, a key brain region involved in emotional processing, plays a crucial role in strengthening memory traces associated with intense emotional experiences. This heightened memory consolidation can lead to a distorted perception of the duration of such events. A traumatic experience, for example, might feel like it stretched out indefinitely, while a joyful occasion might seem to have flown by in an instant. The emotional intensity of the experience, rather than its objective duration, appears to dominate our retrospective evaluation of its temporal span.

Moreover, the narrative structure of our memories further contributes to the subjective nature of time perception. Our memories are not stored as a linear sequence of events but as a network of interconnected episodes, linked through associations, themes, and emotional connections. When we recall a past experience, we often access fragments of memory, stitching them together to create a coherent narrative. This process of narrative construction is inherently creative, introducing the potential for both accurate and inaccurate reconstructions of the sequence and duration of events. The gaps in our memories are frequently filled in with inferences, assumptions, and even fantasies, further blurring the line between objective reality and subjective recollection.

The autobiographical memory system, responsible for our personal life narratives, plays a central role in shaping our overall perception of time's passage. The coherence and consistency of our life narratives influence how we perceive the relationship between past, present, and future. A sense of continuity and meaning in our life story can lead to a perception of time as flowing smoothly and meaningfully, whereas a fragmented or incoherent narrative might result in a feeling of disorientation and a distortion of our temporal experience. This connection between narrative coherence and time perception highlights the active and constructive role of memory in our subjective experience of time.

The phenomenon of "flashbulb memories," vivid and detailed recollections of highly significant events, also demonstrates the complex interaction between memory and time perception. While often assumed to be exceptionally accurate, flashbulb memories are subject to distortions over time, similar to other forms of memory. The initial vividness and confidence associated with these memories might not reflect their actual accuracy. The emotional intensity and societal significance of the event contribute to its memorability, yet this very salience can also lead to reconstructive processes that distort details and even introduce false memories. The perceived clarity and accuracy of these memories, therefore, do not necessarily translate to a more accurate representation of the event's temporal context.

Beyond the individual level, cultural and societal influences also shape our collective understanding and perception of time. Different cultures hold diverse conceptions of time, ranging from linear and cyclical views to those that prioritize the present moment over future planning. These varying perspectives on time's nature influence not only how events are remembered but also how they are experienced in the present. The cultural narratives and shared memories of a community contribute to the collective construction of a temporal landscape, defining what constitutes significant moments, milestones, and even the very structure of time itself. The way a culture conceptualizes and narrates its history significantly impacts its members' individual perceptions of the past, present, and future.

Furthermore, the influence of language on our temporal perception cannot be ignored. The linguistic frameworks we use to describe and conceptualize time inevitably shape our understanding and experience of it. Different languages employ varied grammatical structures and metaphorical expressions to represent time's passage. These linguistic variations can subtly influence how we perceive and remember events, highlighting the intricate connection between language, thought, and temporal experience. The very act of narrating past events in a particular linguistic framework inherently influences how those events are remembered and subsequently perceived.

In summary, the relationship between memory and time perception is a reciprocal and dynamic one. Memory is not simply a storage repository for past experiences but an active reconstructive process that shapes our subjective experience of time's flow. Cognitive biases, emotional salience, narrative construction, cultural influences, and even linguistic frameworks all play significant roles in shaping the way we perceive and remember the past, influencing our perception of temporal duration, sequencing, and even the linearity of time itself. Understanding this complex interplay is crucial to unlocking a deeper understanding of the subjective nature of time and the cognitive mechanisms that underpin our experience of it. The seemingly objective flow of time, as measured by clocks and calendars, is fundamentally reshaped and reinterpreted by the intricate workings of our memory systems, rendering our subjective experience of time a far richer and more complex phenomenon than initially perceived. The study of time perception, therefore, cannot be separated from the study of memory itself; the two are inextricably intertwined in the construction of our subjective reality.

Building upon our understanding of how memory fundamentally shapes our perception of time, we now turn to the fascinating influence of altered states of consciousness. These states, characterized by significant deviations from our typical waking awareness, offer a unique lens through which to examine the subjective experience of temporality. The very structure of time, seemingly immutable in our daily lives, can become fluid and malleable under the influence of altered consciousness, revealing the profound plasticity of our temporal experience.

Meditation, for instance, provides a readily accessible means of exploring this altered state. Through focused attention on the breath, a mantra, or a specific object of contemplation, meditators cultivate a state of heightened awareness, often described as a detached observation of mental processes. Interestingly, time perception frequently undergoes significant transformations during meditative practices. While objective time may continue its relentless march, the subjective experience of duration can dramatically shift. Minutes can feel like hours, or hours can seem to vanish in an instant, depending on the depth and focus of the meditative state. This distortion of temporal perception is not merely a psychological trick; it reflects a fundamental alteration in the brain's processing of temporal information. Neuroimaging studies have demonstrated changes in brainwave patterns during meditation, suggesting a realignment of neural activity that underlies the altered sense of time. The reduced activity in brain regions associated with self-referential thought and the heightened activity in areas related to present-moment awareness contribute to this altered temporal perspective. The sense of "flow," often reported by experienced meditators, is deeply connected to this distorted temporal perception – a seamless merging of the present moment, devoid of the usual distractions and mental chatter that often fragment our experience of time.

Hypnosis, another state of altered consciousness, also provides compelling evidence of the malleability of time perception. Under hypnotic suggestion, individuals can experience significant alterations in their sense of temporal duration and sequencing. Hypnotherapists have reported instances where subjects, under hypnosis, experience past events as if occurring in the present moment, or future events as already experienced. This alteration in temporal perspective highlights the degree to which our conscious awareness can influence the way we perceive and organize temporal information. The power of suggestion, in this context, directly manipulates the brain's internal clock, demonstrating the close relationship between conscious awareness and our subjective experience of time's flow. The suggestibility characteristic of the hypnotic state allows for a direct manipulation of the mental processes involved in constructing our temporal experience. This is not simply a matter of imagination; the reported experiences frequently possess a high degree of emotional intensity and subjective reality, suggesting a genuine alteration in temporal perception rather than merely a fabricated narrative.

The effects of psychoactive substances on time perception are perhaps the most dramatic and widely studied. Many drugs, both recreational and medicinal, significantly alter our experience of temporal duration and sequencing. Substances like LSD, psilocybin, and mescaline are known to induce profound distortions of time perception, with users frequently reporting episodes of time dilation, where moments seem to stretch out indefinitely, or time compression, where extended periods feel like fleeting instants. These alterations are closely linked to the drugs' effects on neurotransmitter systems in the brain, particularly those associated with the processing of sensory information and emotional regulation. The disruption of neurochemical balance can drastically alter the brain's internal temporal markers, leading to the subjective experience of a distorted or fragmented timeline. The intense sensory experiences often associated with these drugs contribute to the perception of time dilation, as the brain struggles to process the overwhelming influx of information. The blurring of boundaries between past, present, and future, a frequent occurrence under the influence of psychedelic substances, further underscores the plasticity of our temporal perception. These experiences, while often described as deeply transformative, highlight the crucial role of neurochemistry in constructing our subjective experience of time. It's crucial to acknowledge the potential risks associated with these substances and to approach their study with caution.

Even everyday experiences can induce subtle alterations in time perception. Engaging activities, such as immersive video games or watching a captivating film, can produce a sense of "flow" – a state where one loses track of time due to intense focus and absorption in the activity. During these periods, the perception of time can be significantly altered, with hours seemingly passing in minutes. This phenomenon mirrors the time distortion observed in meditative or hypnotic states, highlighting the importance of cognitive engagement in shaping our temporal experience. The focused attention required for these activities reduces the processing of time-related cues from the external environment, contributing to the subjective experience of time dilation. This underscores the profound influence of attentional processes on our temporal perception.

The study of altered states of consciousness offers a valuable perspective on the subjective nature of time. The variability of temporal experience observed in these states, ranging from the subtle distortions of meditative states to the dramatic alterations induced by psychoactive drugs, underscores the plasticity of our temporal perception and the complex interplay between brain function, neurochemistry, and consciousness in constructing our subjective experience of time. These findings challenge the notion of a fixed, objective time and reinforce the idea that our experience of time is far more fluid and context-dependent than generally assumed. The investigation of altered states of consciousness continues to shed light on the enigmatic nature of time perception and its inherent subjectivity. It's an area rich in potential for furthering our understanding of consciousness itself and its intimate relationship with our perception of the temporal dimension. The ongoing research into the neural correlates of these altered states is vital in advancing this crucial area of inquiry. Further exploration, employing advanced neuroimaging techniques and rigorous experimental designs, will undoubtedly lead to a more nuanced and comprehensive understanding of the complex interplay between consciousness, neurobiology, and our subjective experience of the passage of time. The seemingly rigid framework of objective time, measured by clocks and calendars, is ultimately refracted and reshaped by the dynamic workings of our minds.

10 Cultural Representations of Non-Linear Time

Moving beyond the subjective distortions of time experienced in altered states of consciousness, we now delve into the rich tapestry of cultural and religious interpretations of time. While our scientific understanding often leans towards a linear, unidirectional model—a continuous progression from past to future—many cultures and spiritual traditions have embraced cyclical conceptions of time, offering alternative frameworks for understanding temporality and its relationship to human existence. These cyclical models challenge the very foundations of our linear worldview, providing fertile ground for exploring non-linearity in its broadest sense.

The concept of cyclical time is deeply rooted in the observation of recurring natural phenomena. The cycle of seasons, the phases of the moon, the daily rise and set of the sun—these repetitive patterns have profoundly shaped human understanding of time throughout history. Rather than viewing time as a linear progression, many ancient cultures perceived it as a continuous cycle of birth, death, and rebirth, mirroring the cyclical processes of nature. This cyclical perspective often intertwines with cosmological views, reflecting the perceived cyclical nature of the universe itself.

Ancient Egyptian cosmology, for example, beautifully illustrates this cyclical understanding of time. Their mythology depicted the sun god Ra's daily journey across the sky, dying each night and being reborn each morning. This daily cycle mirrored their perception of the larger cosmic cycles of creation and destruction, reflecting a cyclical worldview in which time wasn't simply a linear progression but a perpetual recurrence. Similarly, the cyclical nature of the Nile's flooding, essential for their agricultural sustenance, reinforced the importance of cyclical rhythms in their understanding of life and time.

In contrast to the predominantly linear view of time prevalent in Western thought, many indigenous cultures worldwide maintain robust cyclical conceptions of time. For instance, many Native American cultures viewed time not as a straight line but as a circle or a spiral, emphasizing the interconnectedness of past, present, and future. Their understanding of time was deeply embedded in their relationship with the natural world, reflecting the cyclical rhythms of nature and the interconnectedness of all living things. Oral traditions, passed down through generations, preserved their cyclical worldview, embodying a temporal perspective radically different from the linear, historical narratives of Western societies.

Hindu cosmology provides another fascinating example of a cyclical model of time. The concept of *yuga*, a cosmic cycle comprising four ages, each progressively shorter and more degraded than the previous one, represents a grand cyclical process of creation, preservation, and destruction. This cycle, in turn, is part of a larger cycle of *maha-yugas*, further reinforcing the cyclical nature of time within the Hindu worldview. The concept of reincarnation, central to Hinduism, further reinforces this cyclical view, as the soul is believed to repeatedly cycle through birth, death, and rebirth, experiencing different lives and accumulating karma throughout this continuous cycle. The notion of *kala*, time as a cosmic principle, plays a crucial role in underpinning this intricate cyclical framework.

Buddhist cosmology also embraces a cyclical conception of time, albeit with nuances different from the Hindu model. The concept of *samsara*, the cycle of birth, death, and rebirth, is central to Buddhist thought. This cyclical existence is driven by karma, the consequences of past actions that shape future rebirths. The ultimate goal of Buddhist practice is to escape this cycle of suffering and achieve enlightenment – *nirvana* – a state transcending the limitations of cyclical time. However, even the path to nirvana, with its stages of practice and levels of understanding, can be seen as a structured progression within a broader cyclical context.

In contrast to these overtly cyclical systems, some religious and philosophical traditions present a more complex interplay between linear and cyclical elements. Judaism, for example, presents a linear narrative of history, culminating in a messianic age, yet interwoven with cyclical themes, such as the recurring cycles of festivals and agricultural seasons. This juxtaposition of linear and cyclical elements highlights the complexities of temporal experience and suggests that these models are not mutually exclusive but can coexist and inform each other. The cyclical nature of agricultural cycles, inextricably linked to the Jewish faith, emphasizes the enduring importance of cyclical time even within a predominantly linear historical framework.

The implications of these cyclical models for our understanding of non-linear time are profound. They suggest that time is not a unidirectional flow but a multifaceted phenomenon, capable of both linear and cyclical expressions. The cyclical nature of time, as represented in these various cultural and religious contexts, challenges the assumption of a singular, universally applicable model of temporality. It suggests that our perception and understanding of time are deeply shaped by cultural and religious beliefs, highlighting the crucial role of subjective experience in shaping our awareness of time's passage. Furthermore, the cyclical models imply a sense of recurrence, suggesting the possibility of repeating patterns, parallel universes or alternate realities, and the potential for cyclical processes of causality, all challenging the strictly linear causal chains often assumed in scientific models.

The study of cyclical time, therefore, is not merely an academic exercise; it provides valuable insights into the diverse ways humans have conceptualized and experienced the temporal dimension. By understanding these varied perspectives, we can begin to transcend the limitations of our own culturally conditioned perceptions of time and appreciate the richness and complexity of this fundamental aspect of human experience. The exploration of cyclical time opens doors to a deeper understanding of consciousness, the nature of reality, and the potential for profound connections between spirituality, culture, and our subjective experience of the passage of time. This interdisciplinary approach to the study of time invites further research into the neurological underpinnings of these diverse temporal perceptions, promising a more holistic and comprehensive understanding of time itself.

The exploration of cyclical and non-linear time extends beyond the realms of religion and philosophy, finding vibrant expression in art and literature. Here, the fluidity and malleability of time become powerful tools for storytelling, enabling artists and writers to explore the complexities of human experience in ways impossible within a strictly linear framework. Consider, for example, the cyclical nature of many myths and legends. The hero's journey, a common narrative archetype, often involves a descent into the underworld or a transformative trial, followed by a triumphant return—a cyclical structure mirroring the cycles of nature and the human life cycle itself. This cyclical narrative structure provides a framework for exploring themes of death and rebirth, renewal and regeneration, themes which resonate deeply with our understanding of cyclical time. The recurring motifs of seasons, day and night, and the cyclical nature of life itself all contribute to the pervasive presence of cyclical time in narrative structure.

Ancient Greek tragedies, for example, frequently employed cyclical patterns in their plots. The downfall of a hero, often brought about by hubris or fate, could be seen as a cyclical descent, highlighting the inevitability of decline and the cyclical nature of fortune. The chorus, with its recurring refrains and cyclical chants, often reinforced this cyclical sense of time, underscoring the inevitability of fate and the repetitive nature of human suffering. Sophocles' Oedipus Rex, a prime example, demonstrates the cyclical nature of fate and the unavoidable consequences of actions echoing through generations, creating a sense of temporal entanglement that transcends the linear progression of events. The play's tragic climax is not just a singular event but a culmination of actions rooted in the past, inextricably linked to a larger cyclical pattern of fate and retribution.

Moving beyond the classical world, the medieval period witnessed the flourishing of cyclical narratives often intertwined with religious themes. The Divine Comedy, by Dante Alighieri, presents a journey through Hell, Purgatory, and Paradise—a cyclical progression mirroring the soul's journey towards salvation. While the overall narrative structure displays linear elements, the internal structure of each realm, with its concentric circles of punishment or purification, exhibits a cyclical, almost fractal-like, representation of time. Each circle within these realms could represent a stage of spiritual progression, a reflection of cyclical improvement within the larger narrative arc.

The Renaissance witnessed a subtle shift in the portrayal of time in art and literature. While cyclical themes continued to hold significance, the growing influence of humanism and the scientific revolution started to introduce more pronounced linear narratives. However, even during this period, artists often employed cyclical imagery and symbolism to reflect the interplay between earthly and celestial spheres, expressing the cyclical nature of life and death within a larger cosmic framework. The paintings of Hieronymus Bosch, for instance, vividly depict scenes of earthly chaos and heavenly harmony, often arranged in cyclical patterns that highlight the cyclical nature of sin and redemption.

The Baroque period further explored the interplay between linear and cyclical time, often using elaborate allegorical representations to depict the transience of life and the eternal nature of the divine. The memento mori motif, prevalent in Baroque art, serves as a stark reminder of the cyclical nature of life and death, contrasting the ephemeral nature of human existence with the eternal reality of the divine. Paintings depicting scenes of vanitas, emphasizing the fleeting nature of earthly pleasures, reinforced this cyclical perspective by highlighting the inevitability of death as a natural part of the continuous cycle of existence.

The Romantic movement in literature and art embraced cyclical themes, though often with a different emphasis. The Romantic focus on nature and its cyclical rhythms profoundly impacted the depiction of time. Wordsworth's poetry, for instance, is replete with images of nature's cyclical processes—the changing seasons, the cyclical life of plants and animals—reflecting a cyclical perception of time deeply interwoven with the natural world. This connection to nature's cycles brought a renewed appreciation for the cyclical aspects of life, emphasizing the inherent beauty and mystery of both growth and decay.

Modernist literature and art often challenged traditional notions of linear time, employing techniques such as stream-of-consciousness narration, fragmented narratives, and non-linear plot structures to explore the subjective experience of time. James Joyce's Ulysses, for instance, employs a non-linear narrative structure, weaving together various temporal threads, reflecting the complexities of memory and the subjective nature of time. Virginia Woolf's novels similarly explored the non-linearity of human consciousness, using internal monologues and stream-of-consciousness to portray the fluid and fragmented nature of subjective time. These techniques directly challenge the linear progression of time, exploring the interwoven and overlapping experiences that shape an individual's understanding of the passage of time.

Postmodern literature and art further deconstructed traditional conceptions of time, often embracing paradox, fragmentation, and irony. The use of metafiction, self-reflexive narratives, and intertextuality further complicates the linear progression of time, blurring the boundaries between fiction and reality, past and present. These techniques actively resist the notion of a singular, objective timeline, instead offering a kaleidoscopic view of time, a multitude of perspectives and experiences layered and interwoven.

The cinematic arts have also played a significant role in the exploration of non-linear time. Film techniques such as flashbacks, flash-forwards, and dream sequences allow filmmakers to manipulate time, creating non-linear narratives that reflect the subjective experience of time and the fluidity of memory. Christopher Nolan's films, notably Memento and Inception, are masterclasses in the manipulation of time, demonstrating how non-linear storytelling can enhance narrative intrigue and thematic depth. The use of these techniques underscores the capacity of artistic expression to explore the complexities of time beyond the constraints of a linear model.

The cultural representation of time in art and literature reveals a rich tapestry of perceptions, far exceeding the simplistic linear model often assumed in scientific discussions. From the cyclical rhythms of ancient myths to the fragmented narratives of postmodern literature, artistic expressions have consistently engaged with the complexities of time, reflecting and shaping our understanding of this fundamental aspect of human existence. The study of these representations offers invaluable insights into the diversity of human experience, reminding us that our perception of time is deeply embedded within our cultural, religious, and philosophical contexts. It highlights the potential for a more nuanced and holistic understanding of time, acknowledging both its linear progression and its cyclical, even fractal, dimensions, fostering a deeper engagement with the enduring mysteries of time and consciousness. The inherent subjectivity of time's experience, as revealed through artistic expression, offers a compelling counterpoint to the often-rigid, objective models employed in scientific endeavors, enriching our overall comprehension of this fundamental aspect of reality.

The exploration of non-linear time finds particularly compelling expression in the arts of music and dance. These art forms, perhaps more directly than any other, engage with the very fabric of temporal experience, manipulating and shaping our perception of time's flow. While music, at its most basic, can certainly reflect a linear progression – a melody unfolding from beginning to end – its inherent structures and techniques often defy a strictly linear interpretation. The cyclical nature of musical forms, such as rondo, sonata, or even the basic verse-chorus structure of popular music, immediately draws our attention to repetitive patterns and recurring motifs. These repetitions, however, are not simply static reiterations. Each repetition can be subtly altered, building upon the previous iteration, creating a sense of development and transformation within the cyclical framework. This dynamic interplay between repetition and variation is crucial to our understanding of the piece's emotional arc and its overall impact on the listener.

Consider the concept of *ritardando* and *accelerando*, the gradual slowing down or speeding up of tempo. These techniques profoundly affect our perception of time's passage within the musical experience. A *ritardando* at a crucial moment in a piece can stretch time, drawing our attention to the emotional weight of that specific passage. Conversely, an *accelerando* can create a sense of urgency and heightened tension, compressing time and driving the piece towards its climax. The manipulation of tempo becomes a powerful tool for shaping the emotional trajectory of a piece, highlighting the subjective and malleable nature of time within the musical context.

Furthermore, the rhythmic complexity of many musical forms defies easy categorization within a strictly linear timeline. Polyrhythms, where multiple rhythmic patterns occur simultaneously, create a layered and interwoven temporal experience. The listener's attention is constantly shifting between these different rhythmic streams, experiencing time not as a singular, linear progression, but as a multifaceted, simultaneous occurrence. African music, for example, is often characterized by its complex polyrhythmic structures, demonstrating a sophisticated understanding of time that transcends the simple linear model. The simultaneous presence of multiple rhythmic layers creates a temporal texture that defies easy reduction to a single, linear progression, emphasizing the layered and interwoven nature of temporal experience.

The structure of musical pieces itself can be highly non-linear. Composers often employ techniques such as flashbacks or anticipations, mirroring the techniques employed in film to create a sense of temporal disorientation and enhanced narrative depth. The use of thematic transformations, where a musical theme is revisited and modified throughout the piece, can also contribute to a non-linear experience, suggesting a temporal fluidity that transcends the strict sequence of events. The listener is constantly engaging with echoes and allusions to previous sections of the piece, creating a sense of temporal entanglement and a more complex, interwoven understanding of the musical narrative.

Moving to the realm of dance, the interaction between movement, rhythm, and music provides another fascinating lens through which to examine non-linear conceptions of time. Many dance forms, particularly those originating from non-Western cultures, utilize cyclical movements and repetitive patterns to create a sense of timelessness and ritualistic trance. The repetitive nature of these movements, however, does not imply a lack of dynamism or complexity. Subtle variations in posture, speed, or energy can transform a seemingly repetitive movement into a rich tapestry of nuanced expression, demonstrating a complex interplay between repetition and variation, highlighting the multi-layered nature of temporal experience in the art of dance.

Traditional forms such as Indian classical dance, for example, often involve intricate cycles of movements, hand gestures, and rhythmic patterns. These cyclical sequences are not merely decorative, but deeply embedded in the cultural and spiritual context, often reflecting cyclical ideas about life, death, and rebirth. The dancers often use this cyclical structure to build upon previous steps and actions, creating a sense of dynamic expansion within a fundamentally repetitive framework. The structure isn't simply cyclical; it's a cyclical process of growth and transformation, showing how repetition and variation can create a profound, nuanced experience of time.

Contemporary dance often employs more overtly non-linear approaches to the representation of time. The use of improvisation, unpredictable movement sequences, and abrupt changes in tempo can create a disorienting but captivating experience, challenging the viewer's assumptions about the linear progression of time. Dance artists use these non-linear techniques to reflect the fragmentation and fluidity of subjective experience, offering a counterpoint to the often-rigid linear representation of time found in other artistic media. The viewer's perception of time becomes fluid and subjective, mirroring the fragmented and non-linear nature of human consciousness.

Moreover, the interplay between music and dance further complicates the perception of time. The synchronization or asynchronicity between the rhythmic structure of the music and the movements of the dance can dramatically alter our temporal experience. A perfect synchronization can create a sense of harmonious unity, while deliberate asynchronicity can introduce a sense of disjunction and tension, highlighting the contrasting possibilities of our understanding of time's flow. This interplay allows for a dynamic and multifaceted experience of time, transcending the simple linear progression of events to reflect the complexity of human experience.

Consider the concept of *memory* in both music and dance. Musical pieces often evoke memories and associations, transporting the listener to different times and places. Similarly, dance can activate memories and emotions, stimulating a subjective temporal flow that transcends the immediate present moment. The experience of listening to a favorite song from youth, or watching a dance that recalls a specific event, illustrates how these art forms can directly engage with the non-linear and subjective nature of memory, highlighting the complex relationship between personal experience and our perception of time's passage. The temporal linearity is shattered; the past, present, and future intertwine in a profound and deeply personal manner.

In conclusion, the exploration of time in music and dance unveils a rich tapestry of temporal experiences, moving far beyond a simplistic linear model. The cyclical patterns, rhythmic complexities, structural manipulations, and interplay between movement and sound all contribute to a dynamic and multifaceted understanding of time's nature. These art forms, through their inherent structures and techniques, provide powerful means of exploring the subjective, non-linear, and often paradoxical experience of temporality, enriching our appreciation for the multifaceted nature of human perception and consciousness. Through the lens of music and dance, we witness time not as a rigid, predetermined entity, but as a malleable, subjective, and profoundly creative force, reflecting and shaping the very essence of human expression.

The exploration of non-linear time extends beyond the realms of music and dance, finding profound expression in the visual arts of architecture and design. These fields, while seemingly static and concrete, offer remarkably nuanced and multifaceted representations of temporal experience. Unlike music's ephemeral nature, buildings and designs aim for longevity, yet they subtly and sometimes overtly reflect temporal processes, both in their construction and in the way they interact with human experience across generations.

Consider the very act of building itself. The construction of a building unfolds over time, a process of gradual accretion and transformation. Each stage of the building process—from the laying of the foundation to the final touches of interior design—leaves its mark on the finished structure. The building thus becomes a palimpsest, a layered record of its own temporal development. This inherent temporal dimension is often overlooked, yet it profoundly impacts our perception of the building. The visible seams and scars of construction, the subtle variations in materials and techniques used across different phases, can reveal the story of its creation and evolution, giving the building a unique temporal texture. Even the choice of materials, some designed to age gracefully and acquire patina over time, reflects a conscious engagement with the temporal dimension. A weathered stone wall, for instance, tells a story of years spent exposed to the elements, its texture subtly shifting with each passing season.

Furthermore, the design of a building often incorporates symbolic representations of time. Consider the use of clocks, sundials, or other time-measuring devices as architectural elements. These elements not only provide practical functionality, but also serve as visual metaphors for the passage of time, actively engaging the viewer's awareness of their temporal position. The gradual movement of shadows on a sundial, the rhythmic ticking of a clock embedded in a building's façade, transform these functional elements into aesthetic features that resonate with our deep-seated understanding of time's relentless march. The use of such elements within a building transforms the building itself into a temporal instrument, making the passage of time not just an invisible backdrop but an active participant in the architectural experience.

Architectural styles themselves often reflect broader societal understandings of time. Consider the stark, geometric lines of modernist architecture, often associated with a sense of linear progress and technological advancement. This style, prevalent in the early to mid-20th century, reflects a worldview that emphasizes efficiency, functionality, and a forward-looking perspective, with time conceived as a straight line of continuous progress. In contrast, the ornate details and layered complexity of baroque architecture suggest a more cyclical and multi-faceted understanding of time, where history, tradition, and religious beliefs intertwine to create a more richly textured temporal experience. The very weight and presence of baroque architecture, with its opulent detailing and spacious design, suggest a timeless quality, perhaps hinting at a perception of time as something cyclical and enduring, as opposed to the rapid, linear progression suggested by modernist structures.

The spatial arrangement of a building can also subtly manipulate our perception of time. Consider the design of museums and historical sites. These spaces often utilize linear pathways to guide visitors through chronological exhibitions, creating a narrative of historical events that unfolds over time. This deliberate sequencing of exhibits carefully guides visitors through a pre-determined temporal arc, carefully crafted by curators to structure and interpret the past. However, even in such linear environments, the architectural design can offer elements of non-linearity. The presence of unexpected detours, unexpected views, or the sudden exposure to a different stylistic period can disrupt the smooth linear progression, creating temporal interruptions that surprise and intrigue the viewer. These spatial disruptions can create a sense of disorientation and encourage a more nuanced and reflective engagement with the past. Furthermore, the very presence of historical artifacts—ancient statues, faded paintings—can serve as visual reminders of the long passage of time, triggering personal memories and associations that reach beyond the immediate temporal framework of the museum.

In contrast to museums' often-linear approach, consider the design of sacred spaces like temples or cathedrals. These structures often utilize cyclical patterns and repetitive elements to create a sense of timelessness and spiritual transcendence. The repetitive arches of a cathedral, the symmetrical arrangement of columns, or the cyclical pattern of stained glass windows create a visual rhythm that evokes a sense of eternity and spiritual continuity. This repetitive yet varied design echoes the cyclical nature of religious beliefs, mirroring narratives of life, death, and rebirth, and potentially engaging the concept of time as a continuous cycle rather than a straight line. The very stillness and permanence of these spaces invite contemplation, fostering a state of mindfulness where our sense of linear time may recede into the background, allowing for a deeper connection to the present moment and a richer engagement with spiritual themes that often exist beyond our conventional understanding of time.

The use of light and shadow in architecture offers another fascinating dimension to the representation of time. The changing positions of the sun throughout the day and across the seasons dramatically alter the appearance of a building, casting new shadows and highlighting different aspects of its form. This subtle shift in illumination transforms the architecture over time, rendering the building itself as an ever-changing entity. The play of light and shadow is not merely decorative; it actively shapes the experience of the building, encouraging viewers to perceive its beauty in many different temporal contexts. The way light falls on a building can completely change its visual impact, creating a sense of dynamism and constant transformation, emphasizing that even the most static structure is ever-changing over time.

Moreover, urban planning itself can be seen as a form of temporal representation. The layout of a city often reveals its historical development, reflecting layers of past and present. Older neighborhoods, with their winding streets and irregular patterns, may contrast sharply with the grid-like structures of more modern developments. These juxtapositions illustrate the different understandings of time and space that have characterized various periods in the city's history. Furthermore, the city's infrastructure, from transportation networks to public spaces, actively shapes the rhythms of daily life, influencing the flow of human interaction and impacting our subjective experience of time. This dynamic interplay between the built environment and the human experience demonstrates how architectural and urban designs not only reflect temporal processes but also actively shape them.

The interaction between architectural design and human experience further complicates our understanding of time's representation in these fields. Our memories, emotions, and personal associations with specific buildings all contribute to our subjective experience of time within those spaces. A building can become intimately connected to our personal history, acquiring meaning and significance far beyond its purely aesthetic qualities. The passage of time within the memory of a person is never a simple chronological sequence, creating a subjective temporal framework that is uniquely personal. Consider revisiting a childhood home, triggering a rush of memories and emotions that immediately link the present moment to the past, transcending the simple linear sequence of years. This subjective experience, intertwined with the building's objective existence, demonstrates how our perception of time is deeply shaped by our personal relationships with the built environment.

In conclusion, the representation of time in architecture and design is far more nuanced and complex than a simple reflection of linear progression. The building process itself, the symbolic use of time-measuring devices, varying architectural styles, spatial arrangements, the play of light and shadow, and the relationship between the built environment and human experience all contribute to a rich tapestry of temporal representations. By exploring these diverse expressions of time, we can gain a deeper appreciation for the intricate interplay between the built environment and our subjective experience of temporality, recognizing that buildings are not merely static structures, but dynamic entities that actively participate in the unfolding story of human time. They are not simply containers of space, but active participants in shaping our perception of time itself, constantly reminding us of the multifaceted and often non-linear nature of our temporal experience.

The exploration of non-linear time continues into the dynamic world of film and television, where the manipulation of time becomes a powerful tool for storytelling. Unlike the relatively static nature of architecture, cinema and television offer unparalleled flexibility in representing temporal experiences, bending, stretching, and even fracturing the linear progression of time to create compelling narratives and emotional resonance. The very medium itself, with its ability to seamlessly transition between past, present, and future, lends itself naturally to the representation of non-linear temporality.

Consider the ubiquitous use of flashbacks and flashforwards, narrative devices that deliberately disrupt the chronological flow of events to enhance plot development, build suspense, or deepen character development.

The use of flashbacks, for example, allows filmmakers to provide crucial background information, revealing pivotal events from a character's past that illuminate their present actions and motivations. This technique effectively juxtaposes different temporal layers, creating a layered narrative that moves back and forth across time. Consider Christopher Nolan's *Memento*, a film famously structured entirely around fragmented memories, where the narrative unfolds in reverse chronological order. This radical departure from conventional storytelling forces the viewer to actively piece together the narrative, participating in the reconstruction of the protagonist's fragmented past. The experience of watching *Memento* is profoundly non-linear, mirroring the protagonist's own disoriented experience of time and creating a deep sense of suspense and uncertainty.

Similarly, flashforwards offer a glimpse into the future, often used to create anticipation and suspense, or to foreshadow impending events. They disrupt the linear flow of the present, creating a sense of urgency or foreboding. The television series *Lost*, for instance, masterfully employed both flashbacks and flashforwards, intertwining multiple temporal strands to slowly unveil the complex backstory of its diverse characters and the mysteries surrounding their island predicament. The fragmented and interweaving timelines actively challenged the viewer's perception of linearity, demanding active engagement in deciphering the intricate temporal puzzle. This complex temporal structure mirrored the characters' own struggles with their pasts and uncertainties about their futures.

Beyond flashbacks and flashforwards, filmmakers utilize a range of techniques to create non-linear temporal experiences. The use of slow motion and freeze-frames, for instance, can dramatically alter the perception of time, stretching or compressing it to emphasize specific moments. Slow motion, often used in action sequences or moments of intense emotion, allows the viewer to observe details that would otherwise be missed, creating a heightened sense of awareness and emotional impact. Conversely, freeze-frames can halt the narrative completely, giving the viewer time to reflect on a particular image or moment before the story continues. These techniques manipulate the viewer's experience of time within the film's own narrative framework, suggesting that time itself can be pliable and subject to manipulation.

Montage sequences, which present a series of brief, rapidly edited shots, represent another powerful method of manipulating time in film. These sequences can compress a long period of time into a few short seconds, creating a sense of accelerated time and highlighting the passing of days, months, or even years. Such compression enhances the pace of the narrative and selectively emphasize key events, offering a condensed and visually striking representation of time's passage. Conversely, a slow, deliberate montage can also evoke a sense of time stretching and expanding, focusing the viewer's attention on the details of a prolonged process or emotion.

The use of parallel editing, where two or more separate scenes are intercut, also contributes to non-linear storytelling. Parallel editing can create a sense of simultaneity, showing events happening concurrently in different locations or timelines. This technique can build suspense, create dramatic irony, or highlight contrasts between different characters or situations. The famous shower scene in Alfred Hitchcock's *Psycho* masterfully uses parallel editing to heighten the tension and horror, juxtaposing the murder with the seemingly innocent actions of a distant character. This interweaving of simultaneous actions created a disturbing temporal tension that has become iconic in cinema history.

Furthermore, the narrative structure itself can be deployed to represent non-linearity. Nonlinear narratives, where events are presented out of chronological order, are increasingly common in modern film and television. These narratives challenge the viewer's expectations, creating a sense of intrigue and forcing them to actively participate in constructing the story's timeline. The film *Pulp Fiction*, for example, famously utilizes a non-linear structure, jumping between different timelines and perspectives to create a complex and engaging narrative experience. This deliberate disruption of chronological order allowed the film to explore themes of fate, chance, and the interconnectedness of seemingly disparate events in a unique and compelling manner.

Beyond narrative structure, the visual elements of film, such as lighting, color, and set design, can further reinforce the perception of non-linearity.

For instance, a change in lighting or color palette can visually signify a shift in time or location, even without explicit temporal markers. Similarly, set design can contribute to the overall feeling of temporal disorientation, particularly in films that feature temporal shifts or flashbacks. The deliberate use of visual cues reinforces the filmmaker's manipulation of time, subtly guiding the viewer's understanding of the temporal shifts and creating a cohesive visual representation of the non-linear narrative.

Finally, the subjective experience of time, as represented in film, often departs significantly from the objective linear flow. The use of dream sequences, hallucinations, or altered states of consciousness can distort the perception of time within the narrative, creating a subjective and disorienting temporal experience for both the characters and the viewer. Such sequences are often used to explore inner psychological states, conveying the chaotic and fluid nature of subjective time. Films often use these techniques to delve into character psychology, portraying the non-linear and sometimes fragmented nature of memories and internal experiences. The blurring of lines between reality and subjective experience creates a disorienting yet powerful exploration of time's elusive and personalized aspects.

In conclusion, film and television offer a remarkably versatile canvas for representing non-linear time. From the simple use of flashbacks and flashforwards to the more complex manipulation of narrative structure, visual elements, and subjective experience, filmmakers have developed a range of techniques to explore the multifaceted nature of temporality. By skillfully manipulating these tools, filmmakers can create compelling narratives, explore complex themes, and engage viewers in a profound way, enriching our understanding of the fluid and often unpredictable nature of time itself. The ability to manipulate time in film is not just a technical tool but a creative force, fundamentally shaping our engagement with stories and pushing the boundaries of cinematic expression. The dynamic interplay between technical skill and artistic vision underscores the profound power of cinematic representation in shaping our perception of time, moving beyond mere linear chronology to embrace the complexities and paradoxes of temporal experience.

11 The Science of Precognition

The study of precognition, the purported ability to perceive future events, presents a significant challenge to our conventional understanding of time and causality. While dismissed by many within the mainstream scientific community, a substantial body of anecdotal evidence and, to a lesser extent, empirical research suggests the existence of precognitive phenomena. Before delving into the scientific methodologies used to investigate these claims, it's crucial to establish a framework for defining and categorizing the diverse range of precognitive experiences reported. This categorization is far from universally agreed upon, reflecting the nascent and somewhat fragmented state of precognition research. However, a systematic approach is necessary to facilitate meaningful investigation and discussion.

One primary categorization differentiates between precognitive experiences based on their level of clarity and detail. At one end of the spectrum lie vague premonitions, often described as an unsettling feeling of impending doom or a general sense of foreboding, lacking specific details about the nature or timing of the future event. These feelings might manifest as anxiety, unease, or a persistent sense of dread, without any clear imagery or narrative associated with them. For example, an individual might experience a persistent sense of unease about an upcoming trip, a feeling that something might go wrong, without any specific knowledge of what that "something" might be. Such vague premonitions, while potentially indicating a form of precognition, are difficult to empirically validate due to their lack of specificity. The subjective nature of these feelings makes it challenging to establish a clear link between the premonition and a subsequent event.

Conversely, at the other end of the spectrum are highly detailed and specific precognitive experiences, often involving vivid imagery, sensory information, or even a complete narrative sequence of future events. These experiences might include seeing a future event as if it were unfolding before one's eyes, hearing voices predicting a future occurrence, or feeling the emotional impact of an event before it happens. Such detailed precognitive experiences offer greater potential for empirical verification, as their specificity makes it easier to compare the reported precognition to actual future events. For instance, a person might have a dream vividly depicting a car accident involving a specific make and model of car at a particular intersection, and subsequently witness a real-life accident matching the dream's details. The corroboration of such details provides stronger evidence, though it still needs to account for factors like chance and coincidence.

Within the realm of detailed precognitive experiences, further distinctions can be drawn based on the sensory modality involved. Some individuals report primarily visual precognitive experiences, seeing scenes or images of future events. Others may experience auditory precognitions, hearing voices or sounds associated with a future occurrence. Still others may have primarily kinesthetic precognitions, feeling sensations or physical manifestations related to a future event before it actually happens. A complex precognitive experience might incorporate multiple modalities, leading to a rich and multi-layered perception of the future. Consider the case of someone who not only sees a future fire in a building but also hears the screams of people trapped inside and feels the intense heat radiating from the flames before the event unfolds. This multi-sensory experience would represent a more robust and detailed precognitive event, demanding a more rigorous analysis.

Another important aspect of categorizing precognitive experiences lies in differentiating between spontaneous and induced precognitions. Spontaneous precognitions occur without any deliberate attempt to induce or elicit them. They are often unexpected and can vary widely in clarity and detail, ranging from vague feelings to highly specific visions. Induced precognitions, on the other hand, result from deliberate attempts to access or elicit precognitive information, often using techniques such as meditation, visualization, or remote viewing. While both spontaneous and induced precognitions provide valuable data for research, they necessitate different methodological approaches in terms of data collection and analysis. Studies on spontaneous precognitions often rely on retrospective accounts, whereas investigations into induced precognitions typically involve controlled experiments designed to test specific precognitive abilities.

The temporal proximity of the precognitive experience to the future event also constitutes a crucial categorization parameter. Some precognitions occur days, weeks, months, or even years before the predicted event. Others occur just moments before the event, seemingly offering only a very narrow window of precognitive awareness. This temporal dimension significantly influences the challenges of validating the precognition. Long-range precognitions are naturally more difficult to verify due to the vast number of potential variables that could influence the outcome and the difficulty in establishing a clear causal link. Short-range precognitions, however, offer a greater potential for rigorous experimental verification because they limit the possibility of coincidental occurrences.

Furthermore, the emotional valence of the predicted event is relevant. Some precognitive experiences involve positive events, such as winning the lottery or experiencing a joyous reunion, while others relate to negative events, such as accidents, deaths, or disasters. The emotional intensity of the precognition also varies, ranging from mild anxiety to overwhelming terror. This emotional aspect can influence both the nature of the experience and the subsequent recall and reporting of the event. Intensely emotional precognitions, particularly those associated with negative outcomes, are often vividly remembered and reported, but may also be subject to distortion or biases in memory reconstruction.

Beyond these primary categories, a multitude of other factors contribute to the complexity of defining and categorizing precognitive experiences. These include the individual's personality traits, cognitive styles, level of suggestibility, and even cultural background, all of which can influence both the generation and interpretation of precognitive events. Furthermore, the role of chance, coincidence, and confirmation bias cannot be ignored. The human mind is naturally predisposed to find patterns and connections, sometimes interpreting coincidences as evidence of precognition. Rigorous research methods are crucial to separate genuine precognitive experiences from those that can be explained by chance or cognitive biases.

In conclusion, defining and categorizing precognitive experiences is a complex task, demanding a nuanced and multi-faceted approach. A thorough understanding of the various types of precognitive phenomena, ranging from vague premonitions to highly detailed visions, is essential for conducting meaningful research and advancing our understanding of this fascinating and often controversial topic. The development of rigorous and standardized methods for collecting, analyzing, and interpreting data on precognitive experiences is crucial to separating genuine phenomena from chance occurrences and cognitive biases. The field, though fraught with challenges, offers a unique opportunity to examine the limits of our understanding of time, causality, and the nature of consciousness itself. As research methodologies evolve and improve, our ability to understand and potentially even harness the potential of precognition may significantly enhance our comprehension of the universe and our place within it.

The exploration of precognition necessitates a rigorous examination of empirical studies attempting to scientifically validate the phenomenon. While anecdotal evidence abounds, the scientific community demands demonstrable, repeatable results under controlled conditions. The challenge lies in designing experiments that minimize confounding variables and account for potential biases inherent in both the subjects and the researchers. Several approaches have been employed, each with its own strengths and limitations, and a critical assessment of these studies is crucial for evaluating the current state of precognition research.

One prominent approach involves the use of ganzfeld experiments. These experiments aim to create a state of sensory deprivation in the participant, theoretically making them more susceptible to receiving precognitive information. The participant is placed in a dimly lit, soundproofed room, wearing halved ping-pong balls over their eyes and listening to white noise. Meanwhile, a "sender" concentrates on a randomly selected target image or video. After a period of sensory deprivation, the participant is presented with four images or descriptions, one of which is the target. Their task is to identify the target image. While some ganzfeld studies have reported statistically significant results, suggesting a higher-than-chance rate of successful target identification, these findings remain highly controversial. Critics point to methodological flaws, such as the lack of proper randomization procedures, potential biases in data selection and analysis, and the difficulty in replicating results across different labs and researchers. The subtle nature of the effect, the challenges in controlling for chance, and the susceptibility to experimenter bias significantly hinder the acceptance of ganzfeld studies as definitive proof of precognition. Moreover, the underlying mechanism by which sensory deprivation might enhance precognitive abilities remains unexplained, leaving the interpretation of positive results open to numerous alternative explanations.

Another experimental paradigm frequently employed in precognition research involves forced-choice precognition experiments. In these studies, participants are asked to predict a future random event, such as the outcome of a coin toss or the selection of a random number. The critical aspect is that the random event occurs *after* the participant makes their prediction. Statistical analysis then determines whether the participants' predictions exceed chance levels. While some studies using this methodology have reported positive results, the effect sizes are typically small and often fail to replicate across independent studies. Furthermore, concerns regarding subtle biases in experimental design, data analysis, and participant selection continue to plague this area of research. The possibility of unconscious cues, experimenter bias, or statistical artifacts influencing the results makes it difficult to definitively attribute positive findings to genuine precognition. The necessity for large sample sizes to achieve statistical significance also presents logistical and financial challenges, limiting the scope and replicability of these experiments.

Retrospective studies, analyzing past events for evidence of precognitive experiences, represent another approach. These studies often involve analyzing diaries, personal accounts, or news archives to identify instances where individuals accurately predicted future events. While such studies can provide valuable qualitative data and intriguing examples, they are notoriously susceptible to various biases. Memory distortion, confirmation bias, and the difficulty in establishing a clear temporal sequence between the precognitive experience and the subsequent event all significantly compromise the reliability of this method. The subjective nature of interpreting past events and the inherent challenges of establishing causality make retrospective studies inherently weak in providing conclusive evidence for precognition. While these studies may generate interesting hypotheses and identify potential avenues for further research, they cannot stand alone as definitive proof of precognition.

Remote viewing, a technique involving attempts to perceive a distant or shielded location or object, has also been explored in connection with precognition. Proponents argue that the ability to remotely view a location implicitly involves some form of precognitive ability, as the viewer is perceiving information before it is directly available to them through conventional sensory means. However, similar to other methodologies, remote viewing studies have been criticized for methodological flaws, including unclear judging criteria, lack of rigorous controls, and a tendency towards confirmation bias. The inconsistent results and the difficulty in replicating positive findings across different research groups have prevented the mainstream acceptance of remote viewing as a valid demonstration of precognition. The subjective nature of interpretation, the potential for subtle cues, and the difficulty in ruling out chance occurrences significantly weaken the conclusions drawn from remote viewing studies.

Meta-analyses of existing precognition studies have been undertaken, attempting to combine the results of multiple studies to identify overall trends and assess the strength of evidence. However, these meta-analyses are often hampered by the heterogeneity of the methodologies used in the individual studies, making it difficult to draw robust conclusions. The inconsistencies in experimental designs, the variation in effect sizes, and the limitations of the individual studies often limit the power of meta-analyses to provide definitive evidence for or against precognition. Moreover, publication bias, the tendency for studies with positive results to be published more readily than those with null or negative results, can significantly skew the outcome of a meta-analysis, leading to an overestimation of the true effect size of precognition.

In summary, the empirical evidence for precognition remains inconclusive and fragmented. While several studies have reported statistically significant results suggestive of precognitive abilities, these findings are often contested due to methodological limitations, concerns about replicability, and the possibility of alternative explanations. The inherent challenges of designing rigorous experiments to test precognitive phenomena, the subtle nature of the effects, and the potential for biases in both the participants and the researchers all contribute to the ongoing debate. Future research must prioritize rigorous methodology, including rigorous control for bias, clear and reproducible experimental designs, and large-scale studies to achieve sufficient statistical power. Furthermore, a deeper understanding of the potential underlying mechanisms of precognition, if it indeed exists, is essential for advancing this field of study. Only through the development and application of more robust and sophisticated research methods can we hope to arrive at a more conclusive understanding of the scientific validity of precognition.

The inconclusive nature of empirical studies on precognition necessitates a deeper dive into the potential psychological and neurological underpinnings of such experiences. While the existence of precognition itself remains debated, exploring the cognitive and neural mechanisms associated with reported precognitive events can offer valuable insights, regardless of whether those events reflect genuine precognition or alternative explanations. This investigation requires a nuanced approach, acknowledging the limitations of current research while exploring promising avenues for future investigation.

One critical area of exploration involves cognitive biases. Confirmation bias, the tendency to favor information confirming pre-existing beliefs, plays a significant role in interpreting ambiguous experiences. Individuals who believe in precognition might be more likely to remember instances seemingly supporting this belief while dismissing contradictory evidence. This selective recall can artificially inflate the perceived frequency of precognitive experiences. Similarly, the availability heuristic, where readily available information is overestimated in frequency, might lead individuals to overestimate the likelihood of precognitive events. The vividness of a seemingly precognitive experience, irrespective of its actual predictive accuracy, makes it more memorable and readily accessible, thus reinforcing the belief in precognition. Furthermore, hindsight bias, the tendency to perceive past events as more predictable after they have occurred, can create a false impression of precognitive accuracy. An event that seemed surprising beforehand may appear inevitable in retrospect, leading to a misattribution of foresight.

Beyond cognitive biases, altered states of consciousness (ASC) may also be relevant. Many reported precognitive experiences occur during periods of heightened suggestibility or reduced critical thinking, such as during dreams, meditation, or under the influence of certain substances. ASC may lower the threshold for accepting unusual sensory inputs or unusual associations, potentially leading to experiences interpreted as precognitive. The reduced filtering of sensory information characteristic of ASC might allow for the processing of weak or ambiguous signals that would normally be ignored, creating a sense of knowing something before it happens, even if this "knowing" is coincidental or based on unconscious inference. The exploration of ASC and precognition necessitates a careful consideration of the various types of ASC, their neurobiological correlates, and their potential to influence cognitive processing in ways that could lead to the misinterpretation of random events as precognitive experiences.

Neurological investigations offer further avenues for understanding the potential mechanisms of precognition. While there is no single "precognition center" in the brain, research on brain activity during precognitive tasks might reveal specific patterns or neural correlates associated with these experiences. Neuroimaging techniques like EEG (electroencephalography) and fMRI (functional magnetic resonance imaging) could be employed to investigate brain activity during tasks designed to elicit precognitive responses. Analysis of EEG data might reveal distinct brainwave patterns associated with reported precognitive experiences, potentially shedding light on the neurophysiological states linked to such phenomena. Similarly, fMRI studies could identify specific brain regions showing increased activity during precognitive tasks, providing insights into the neural networks involved. Such research, however, is challenging due to the difficulty of reliably inducing and measuring precognitive events under controlled experimental conditions. The subtle nature of the presumed effect and the potential confounding factors need careful attention in experimental design and data analysis.

The potential role of specific brain regions and networks should be explored. For example, the default mode network (DMN), active during mind-wandering and self-referential thought, might be implicated in generating intuitive hunches or feelings that are later interpreted as precognitive. The DMN's involvement in generating spontaneous thoughts and associations could potentially lead to the creation of seemingly accurate predictions, even if they are based on chance or unconscious inference. Furthermore, brain areas associated with memory consolidation and emotional processing might also play a role, as strong emotions are frequently associated with precognitive experiences. The limbic system, particularly the amygdala and hippocampus, might contribute to the vividness and memorability of such experiences, influencing subsequent recall and interpretation. Exploring these potential neural correlates demands sophisticated neuroimaging techniques and rigorous experimental design to minimize confounding factors and maximize the validity of the findings.

Further investigations should focus on the temporal dynamics of precognition. Does the feeling of precognition occur instantaneously, or does it develop over time? If the latter is true, the study of the temporal sequence of neural events leading to a reported precognitive experience might provide crucial clues. This could involve the use of high-temporal-resolution neuroimaging techniques like MEG (magnetoencephalography) to track the progression of brain activity associated with precognitive experiences. Analyzing the temporal relationship between neural activity and the subsequent event could provide insights into the predictive accuracy of these experiences, even if that accuracy is statistically slight. Such research should not only focus on the moment of the "precognitive experience" but also consider the potential for unconscious processes to influence subsequent behaviour, thereby shaping the likelihood of an outcome that aligns with the perceived prediction.

The study of exceptional individuals who consistently report high rates of successful precognitive experiences – sometimes termed "psi-gifted" individuals – may provide further insights. While this group remains a challenge to study due to the difficulty of reliably identifying and recruiting participants who genuinely exhibit such abilities, their reported experiences offer a potential window into the psychological and neurological mechanisms underlying precognitive phenomena. It is crucial, however, to approach studies of such individuals with caution, acknowledging the potential for bias and the need for rigorous control measures. Independent verification of reported events, along with thorough psychological and neurological assessments, are necessary to ensure the reliability of any findings from studies of "psi-gifted" individuals. Such studies require careful consideration of ethical issues surrounding participant selection and data interpretation.

In conclusion, unraveling the psychological and neurological underpinnings of precognition necessitates a multidisciplinary approach that combines cognitive psychology, neuroscience, and potentially, parapsychology, acknowledging both the challenges and potential breakthroughs within this complex field. Although the empirical evidence supporting precognition remains limited and controversial, the exploration of the cognitive biases, altered states of consciousness, and neural activity associated with reported precognitive experiences can advance our understanding of human perception, cognition, and the mind's potential for processing information in ways that currently elude our scientific comprehension. Future research focused on rigorous methodology, large-scale studies, and multi-faceted investigations could potentially yield significant advances in our understanding of this enigmatic phenomenon. The cautious yet open-minded exploration of these complex phenomena is crucial for advancing our scientific understanding of the human mind and its possible interaction with reality in ways beyond our current explanatory frameworks.

The persistent skepticism surrounding precognition stems from several interconnected factors, primarily the lack of robust, replicable experimental evidence and the inherent difficulties in designing studies that adequately control for chance occurrences and cognitive biases. Skeptics often point to the prevalence of coincidence and the human tendency to selectively remember instances that seem to confirm precognitive abilities while ignoring instances that contradict them. This confirmation bias, a well-documented cognitive phenomenon, can significantly distort the perceived accuracy of precognitive experiences. A seemingly prophetic dream, for example, might be vividly recalled if the predicted event occurs, while numerous other dreams containing inaccurate predictions are quickly forgotten. This selective memory creates an inflated impression of precognitive accuracy.

Furthermore, the issue of statistical significance is often raised as a crucial point of contention. Many precognition studies report results that are statistically insignificant, meaning that the observed outcomes could easily be attributed to chance alone. Even if a study reports a statistically significant result, replicating that finding in subsequent studies is frequently problematic, leading to doubts about the robustness and generalizability of the initial results. The absence of a consistently replicable effect is a significant hurdle in establishing precognition as a genuine phenomenon. The small effect sizes reported in many studies, even those showing statistical significance, also contribute to the skepticism. Small effects, while statistically significant, may not be practically significant and could represent a weak or unreliable effect.

Methodological limitations within precognition research further exacerbate the skepticism. Many studies suffer from issues such as poor experimental design, inadequate controls, lack of blinding (where participants and experimenters are unaware of the experimental conditions), and insufficient sample sizes. The lack of standardized protocols for assessing and measuring precognitive experiences also hampers the comparability and replicability of results across different studies. The subjective nature of precognitive experiences, relying heavily on participant reports, introduces further challenges in ensuring the objectivity and validity of the findings. The inherent difficulty in objectively defining and measuring precognition contributes to the difficulty of designing rigorous and unbiased studies. The operational definition of precognition itself is a source of debate, impacting the design and interpretation of research.

Another significant criticism revolves around the potential for fraud or experimenter bias. The subtle nature of precognitive effects makes them susceptible to manipulation, either intentional or unintentional. Subtle cues from the experimenter, even unconscious ones, could influence participant responses, leading to artificially inflated results. Similarly, participants might unconsciously bias their responses to conform to expectations, either due to suggestibility or a desire to please the experimenter. The lack of double-blind procedures, where neither the participant nor the experimenter knows the experimental conditions, increases the vulnerability to such biases. Stricter control measures, including randomization, blinding, and independent verification of results, are crucial to mitigate these concerns and bolster the credibility of the research.

The role of chance and coincidence in seemingly precognitive experiences also fuels skepticism. Given the vast number of events occurring every day, some coincidences are inevitable. The human brain is adept at finding patterns and connections, even where none exist. This apophenia – the tendency to perceive meaningful connections between unrelated events

– can lead to the misinterpretation of random occurrences as precognitive events. The human mind's capacity for pattern recognition and its propensity for constructing narratives can easily create a sense of precognition where none actually exists. Statistical analysis, taking into account the probability of coincidental occurrences, is essential in evaluating the validity of claimed precognitive experiences.

Furthermore, psychological factors beyond cognitive biases contribute to the controversies. For instance, anxiety and stress can enhance suggestibility, making individuals more susceptible to believing in precognitive experiences even in the absence of any actual predictive ability. The desire for certainty and control in an uncertain world can also contribute to the acceptance of precognitive explanations. The allure of the supernatural, the search for meaning and purpose, and the comfort in believing in a future that is somewhat predictable can significantly influence both the perception and reporting of precognitive experiences. Understanding these psychological underpinnings is crucial in interpreting precognition research findings.

The critiques extend beyond experimental methodology. Philosophical objections question the very possibility of precognition. These objections often center on the concept of causality and the implications of precognition for our understanding of time and determinism. If precognition is genuine, it would seem to challenge our basic understanding of cause and effect, raising profound questions about free will and the predictability of future events. The apparent paradox of influencing past events through future knowledge also presents a major philosophical stumbling block. These are complex issues that require careful consideration, transcending the purely empirical questions of experimental design and data analysis.

Despite these significant criticisms and challenges, the exploration of precognition should not be dismissed entirely. The fact that such experiences are widely reported across cultures and throughout history warrants continued, albeit cautious, investigation. The potential for refining research methodologies, employing advanced neuroimaging techniques, and developing more sophisticated statistical analyses could lead to a more nuanced understanding of the phenomenon, irrespective of whether it ultimately confirms the existence of genuine precognition. The ongoing challenge is to design studies that can rigorously address the methodological and philosophical concerns raised by skeptics, potentially through interdisciplinary collaborations bringing together expertise in psychology, neuroscience, statistics, and philosophy. Only through rigorous, replicable research can we hope to gain a clearer understanding of the nature and extent of precognitive experiences and whether they reflect a fundamental aspect of human consciousness or simply a collection of cognitive biases and coincidences.

The path forward in precognition research necessitates a multifaceted approach, transcending the limitations of past studies. One crucial area is the development of more sophisticated experimental designs. Current methodologies often rely on relatively simple tasks, such as predicting random events or future stock market fluctuations. Future research should explore more complex scenarios that better reflect the nuanced and often ambiguous nature of precognitive experiences. For example, studies could focus on predicting complex sequences of events, utilizing scenarios that require integration of multiple sensory inputs and contextual understanding, moving beyond simple binary choices. This complexity would make it more difficult for chance or cognitive biases to account for accurate predictions. The development of more naturalistic paradigms—situations that mirror real-world experiences—is essential. Instead of laboratory-based tasks, researchers might investigate precognitive experiences in the context of everyday life, using ecological momentary assessment methods to collect data in real-time. This approach could yield more authentic and ecologically valid data, minimizing the artificiality inherent in many previous studies.

Furthermore, incorporating advanced technologies could significantly enhance the precision and objectivity of precognition research. Neuroimaging techniques, such as fMRI and EEG, offer the potential to identify neural correlates of precognitive experiences. By monitoring brain activity during precognitive tasks, researchers could potentially identify specific brain regions or patterns of activity associated with accurate predictions. This neurophysiological approach could provide more objective evidence, reducing reliance on subjective self-reports and bolstering the credibility of precognition findings. The application of eye-tracking technology could also prove valuable in detecting subtle cues or changes in gaze patterns that might indicate precognitive processing. This could help elucidate the cognitive mechanisms involved in precognition. Moreover, advanced statistical methods, such as Bayesian analysis and machine learning algorithms, could provide more robust and sophisticated ways to analyze the often-noisy data obtained in precognition studies. These advanced tools are better equipped to handle complex data sets and identify subtle patterns that might be missed by traditional statistical techniques.

Another crucial avenue for future progress lies in interdisciplinary collaboration. Precognition research should not be confined to the realm of parapsychology alone. A truly comprehensive understanding requires the integration of insights from various disciplines. Neuroscience, cognitive psychology, and computer science could provide valuable theoretical frameworks and methodological tools. For instance, cognitive scientists could help refine the measurement of precognitive experiences, develop validated scales to assess the accuracy and confidence of predictions, and identify potential cognitive biases that could influence the results. Neuroscientists could investigate the neural mechanisms underlying precognitive experiences, while computer scientists could develop sophisticated algorithms to analyze the vast amounts of data generated in such studies. The collaboration of philosophers is also critical to address the fundamental conceptual challenges surrounding precognition, clarifying the nature of time, causality, and the implications for free will. Interdisciplinary dialogues could facilitate the development of more rigorous experimental designs, more nuanced interpretations of findings, and a deeper understanding of the potential implications of precognition.

The challenge of replicability, a cornerstone of scientific validity, remains paramount. Future studies must prioritize rigor and transparency in their methods. Detailed descriptions of experimental protocols, including participant selection criteria, data collection procedures, and statistical analyses, are essential for enhancing replicability. Pre-registration of studies, outlining the hypotheses and methodology before data collection, is a key step in preventing publication bias and promoting transparency. Open-access data sharing, allowing independent researchers to scrutinize the raw data and conduct their own analyses, is another critical component of enhancing the scientific rigor of precognition research. The use of standardized protocols across different laboratories would also greatly enhance comparability and the ability to confirm findings. Independent replication efforts by different research teams are crucial for establishing the robustness of any observed effects.

Beyond methodological improvements, a more comprehensive understanding of individual differences is needed. Not all individuals report the same level of precognitive experiences, suggesting that there might be variations in susceptibility or ability. Future research should investigate potential factors influencing individual differences, such as personality traits, cognitive styles, and genetic predispositions. This could involve longitudinal studies, tracking individuals over time to assess changes in precognitive abilities and identify potential predictors. Studying the potential relationship between mindfulness practices or altered states of consciousness and precognitive experiences could also prove fruitful. These investigations would help explain why certain individuals appear to possess stronger precognitive abilities than others.

Further, the exploration of potential mechanisms underlying precognition requires a more nuanced approach. While some theories propose that precognition reflects a form of extrasensory perception, others suggest that it might arise from subtle unconscious biases, enhanced pattern recognition, or even quantum entanglement. Future research should critically evaluate these competing explanations and identify the most plausible mechanisms, drawing on insights from various scientific disciplines. This would involve developing testable hypotheses and designing experiments specifically tailored to differentiate between these competing explanatory models. The integration of different theoretical perspectives, considering both the psychological and potentially physical aspects of precognition, is crucial for developing a comprehensive understanding.

Finally, a crucial aspect of future research involves addressing the ethical implications of precognition. If precognitive abilities are demonstrated to be genuine, this would have profound implications for various fields, including medicine, finance, and national security. Careful consideration is required to assess the potential risks and benefits of harnessing such abilities. Ethical guidelines and regulations would be needed to prevent misuse and ensure responsible application of any validated precognitive technologies. The ethical discussion should encompass the potential for psychological harm, the societal consequences of manipulating future events, and the equitable distribution of any benefits derived from precognitive research.

In conclusion, advancing the scientific understanding of precognition requires a multipronged strategy. This includes developing innovative experimental designs, employing advanced technologies, fostering interdisciplinary collaborations, addressing the challenge of replicability, investigating individual differences, exploring potential mechanisms, and critically considering ethical implications. By adopting a more rigorous and comprehensive approach, future research has the potential to shed light on this fascinating and controversial phenomenon, whether it confirms precognition as a genuine ability or ultimately illuminates the workings of the human mind and its propensity for pattern recognition and subjective interpretation. The journey towards understanding precognition, though fraught with challenges, promises to be profoundly rewarding in enriching our understanding of consciousness and the nature of reality itself.

12 The Science of Déjà Vu

The elusive nature of déjà vu, that unsettling feeling of having already experienced the present moment, has long captivated researchers. While its precise cause remains a mystery, significant progress has been made in understanding the potential neurological mechanisms underpinning this intriguing phenomenon. One prominent theory centers on the idea of a memory processing malfunction within the brain. Our brains are constantly encoding and retrieving memories, a complex process involving multiple brain regions working in concert. Déjà vu might arise from a glitch in this intricate system, a temporary disruption in the normal flow of information between memory encoding and retrieval systems.

Imagine the brain as a vast library, meticulously organized and cross-referenced. In the normal course of events, when we encounter a new experience, the brain carefully catalogs the sensory details, emotional context, and associated thoughts. This newly encoded memory is then stored in its appropriate location within the "library." Later, when we encounter a similar experience, the brain retrieves the relevant memory, providing a sense of familiarity. However, in cases of déjà vu, this retrieval process might be slightly "off." The brain might inadvertently access a similar, albeit not identical, memory from the past, leading to the false impression that the current experience has been lived before. This "false memory" is so convincing that it generates the potent sense of familiarity associated with déjà vu.

Several brain regions are implicated in this potential memory-related dysfunction. The hippocampus, a seahorse-shaped structure deep within the brain, plays a crucial role in forming new memories. Damage to the hippocampus, or even subtle disruptions in its function, could interfere with the precise encoding and retrieval of memories, potentially contributing to déjà vu. Similarly, the parahippocampal gyrus, a cortical region closely associated with the hippocampus, is involved in processing spatial information and contextual details of memories. A momentary lapse in its function could lead to a blurring of the boundaries between past and present, resulting in the misattribution of familiarity to a novel experience.

Furthermore, the prefrontal cortex, responsible for higher-level cognitive functions such as decision-making and working memory, might be involved. The prefrontal cortex acts as an executive control center, overseeing the various cognitive processes involved in memory. If its function is temporarily impaired, it might fail to properly distinguish between new and old memories, leading to the subjective experience of déjà vu. The amygdala, the brain's emotional center, also plays a role. While not directly involved in memory encoding and retrieval, the amygdala's processing of emotional information might contribute to the intensity and vividness of the déjà vu experience. A strong emotional association with a similar past memory could amplify the sense of familiarity, making the déjà vu experience more profound.

Another compelling explanation focuses on neural misfirings or disruptions in brainwave patterns. The brain's electrical activity constantly fluctuates, generating various brainwave patterns associated with different mental states. Déjà vu might be triggered by a temporary disruption in these patterns, a brief synchronization of neural activity that mimics the pattern associated with a similar past experience. This could lead to a spontaneous activation of a related memory network, producing the feeling of familiarity without the conscious retrieval of the specific memory itself.

Research using electroencephalography (EEG) has shown altered brainwave patterns during experiences of déjà vu. While the exact nature of these alterations remains unclear, they could reflect a temporary change in the way different brain regions communicate with each other. This communication breakdown might lead to a false sense of familiarity, generating the subjective experience of déjà vu. The intricate interplay between different brain regions, and the subtle fluctuations in their electrical activity, highlights the complexity of the neural mechanisms underlying this phenomenon.

The temporal lobes, located on the sides of the brain, have been implicated in many studies of déjà vu. The temporal lobes are critical for memory consolidation and retrieval, and damage to these areas, whether from trauma, epilepsy, or other neurological conditions, can frequently lead to an increased incidence of déjà vu experiences. This observation suggests that a dysfunction in the temporal lobe circuitry might be a crucial factor in the genesis of déjà vu.

Interestingly, déjà vu is not limited to neurological disorders. Healthy individuals experience it too, although typically much less frequently. This suggests that the underlying mechanisms might involve temporary fluctuations in brain activity rather than permanent structural damage. The fact that stress, sleep deprivation, and certain drugs can increase the frequency of déjà vu episodes lends further support to this idea. These factors can induce temporary disruptions in brain function, potentially triggering the neural misfirings or memory processing glitches responsible for déjà vu.

The phenomenon of déjà vu is further complicated by the difficulty in systematically studying it. It is a subjective experience, making it challenging to objectively measure and quantify. The fleeting and unpredictable nature of déjà vu makes it hard to design controlled experiments that reliably elicit the experience. Moreover, the reported instances of déjà vu often lack the detail needed to definitively confirm or refute proposed theories. The individual's subjective interpretation of the event, influenced by personal biases and expectations, can significantly affect their reported experience.

Despite these challenges, advances in neuroimaging techniques offer new possibilities for studying déjà vu. Functional magnetic resonance imaging (fMRI), which measures brain activity through changes in blood flow, allows researchers to examine the neural correlates of déjà vu in greater detail. Studies using fMRI have provided some evidence supporting the involvement of the hippocampus and other memory-related brain regions in déjà vu experiences. However, more research is needed to fully understand the complex interplay of these regions and to pinpoint the precise neural mechanisms involved.

Future research should focus on developing more sophisticated experimental paradigms that can reliably induce déjà vu in a controlled setting. This could involve virtual reality simulations that recreate situations likely to trigger déjà vu or the use of carefully designed stimuli that selectively activate specific brain regions. The integration of different neuroimaging techniques, such as EEG and fMRI, could provide a more comprehensive picture of the neural activity patterns associated with déjà vu.

Moreover, a deeper understanding of individual differences in susceptibility to déjà vu is crucial. Some people report experiencing déjà vu frequently, while others rarely or never encounter it. Exploring potential factors contributing to this variability, such as genetic predispositions, personality traits, or differences in brain structure, could reveal further clues about the underlying mechanisms. Studies investigating the potential link between déjà vu and other cognitive phenomena, such as memory distortions and illusions of familiarity, could also be valuable.

In conclusion, while the exact neurological mechanisms causing déjà vu remain elusive, accumulating evidence points toward a combination of memory processing glitches and transient disruptions in brainwave patterns. The hippocampus, parahippocampal gyrus, prefrontal cortex, and amygdala are likely involved, potentially due to temporary disruptions in their function or communication. Further research using advanced neuroimaging techniques and sophisticated experimental paradigms is needed to unravel the intricate neural processes that underlie this fascinating and enigmatic phenomenon, furthering our understanding of the complex workings of human memory and consciousness. The pursuit of knowledge in this field is not merely an academic exercise; it offers insights into the very nature of how our brains construct our experience of reality, even when it subtly deceives us.

Beyond the purely neurological interpretations, the psychological dimensions of déjà vu offer a compelling avenue for understanding this intriguing phenomenon. While neurological theories focus on the "hardware" of the brain, psychological explanations delve into the "software"—the cognitive processes that shape our perception and interpretation of experience. One prominent psychological perspective centers on the concept of memory distortion. Our memories are not static recordings of events; they are dynamic constructs, subject to alteration and embellishment over time. The act of remembering is itself a process of reconstruction, influenced by our current emotional state, expectations, and pre-existing knowledge.

Déjà vu, from this perspective, could be interpreted as a form of memory illusion, a mistaken attribution of familiarity to a novel experience. This misattribution arises not from a neurological malfunction per se, but rather from a cognitive error in the process of memory retrieval and comparison. The brain might encounter a new situation that evokes a faint resemblance to a past experience, a resemblance strong enough to trigger a feeling of familiarity but not strong enough to access the original memory consciously. This "echo" of a past memory, however faint, is sufficient to generate the feeling of "having been here before."

Consider, for example, walking into a new restaurant. The layout, the décor, even the subtle aroma might unconsciously trigger a feeling of familiarity, reminiscent of another place, perhaps a long-forgotten vacation or childhood home. The brain, struggling to locate the source of this feeling, might mistakenly attribute the familiarity to the current context, resulting in a fleeting experience of déjà vu. The lack of explicit recall of the original memory is crucial here; if you could clearly remember the similar place, it wouldn't be déjà vu.

This perspective aligns well with the observation that déjà vu often occurs in ambiguous or unfamiliar situations. When the brain faces a novel context, the process of constructing a mental representation becomes more challenging and susceptible to errors. In such scenarios, the brain might be more likely to misattribute a subtle resemblance to a past experience as a true replication of that experience, leading to the experience of déjà vu. Furthermore, this aligns with the frequently reported feeling that déjà vu is accompanied by an absence of specific details. It's not a complete reliving of a past moment, but rather a faint impression of familiarity lacking sharp outlines.

Another compelling psychological explanation focuses on the concept of "familiarity illusion." This refers to the feeling of knowing something without consciously remembering its source. This feeling can manifest in various forms, such as the tip-of-the-tongue phenomenon, where you know you know the answer but can't recall it, or the experience of hearing a song and feeling as if you have already heard it before even though you know you haven't. Déjà vu might be a particularly vivid and potent manifestation of this familiarity illusion, a false sense of knowing applied to a specific spatial or temporal context.

Several cognitive processes could contribute to this familiarity illusion. One is the brain's tendency to create associations between seemingly unrelated pieces of information. A seemingly new situation might share subtle similarities with a past event stored in memory, creating a subconscious link that generates the feeling of familiarity. This association might be strengthened by factors such as emotional context or sensory details, thereby enhancing the intensity of the déjà vu experience. The unconscious nature of this process emphasizes the complex interplay of different cognitive systems that operate outside of conscious awareness.

Furthermore, the brain's capacity for pattern recognition plays a significant role. We are wired to identify patterns and regularities in our environment. This allows us to make predictions and navigate our surroundings efficiently. However, this remarkable ability can also lead to misinterpretations. The brain might misinterpret a new experience as a previously encountered pattern, even if the similarity is superficial. This misinterpretation triggers a sense of familiarity, manifesting as déjà vu.

The interplay of attention and memory also influences the likelihood of experiencing déjà vu. Situations that capture our attention strongly tend to be encoded more vividly in memory. If the attentional focus is intense, the brain might create a strong association between the elements of a new situation and a similar past experience. This strong association might lead to a more compelling sense of familiarity, even though the actual similarity is minimal. Similarly, memory consolidation, the process by which the brain strengthens memories over time, plays a role. Newly encoded memories are particularly vulnerable to distortions, and the process of memory consolidation could contribute to the illusion of familiarity associated with déjà vu.

The psychological aspects of déjà vu are further complicated by individual differences. Some individuals report experiencing it frequently, while others rarely or never do. This variation highlights the influence of factors like personality traits, coping mechanisms, and even cultural background. For example, individuals with high levels of anxiety or those prone to rumination might be more susceptible to experiencing déjà vu, potentially due to a heightened sensitivity to memory distortions or a tendency to over-analyze sensory information.

In contrast, individuals with a more relaxed and less analytical mindset might be less prone to such experiences. Furthermore, cultural beliefs and expectations surrounding the phenomenon could influence the frequency with which it is reported. A culture that embraces a belief in precognitive abilities or past lives might encourage individuals to attribute déjà vu to paranormal explanations and thus report it more frequently.

In conclusion, psychological theories offer valuable insights into the subjective experience of déjà vu, complementing the neurological interpretations. Memory distortions, familiarity illusions, and the dynamic interplay of cognitive processes all contribute to the creation of this potent sense of familiarity. Individual differences in cognitive style, personality traits, and cultural context further modulate the likelihood and interpretation of déjà vu experiences. A comprehensive understanding of this phenomenon requires integrating both neurological and psychological perspectives, recognizing the intricate interplay of brain function and cognitive processes in shaping our subjective reality. The mystery of déjà vu, therefore, remains a captivating example of how the human mind, both in its physiological and cognitive complexity, shapes our perception of reality, even when that perception subtly misrepresents the truth. Further research involving interdisciplinary collaborations between neuroscientists, psychologists, and cognitive scientists will be essential to fully unveil the secrets of this ubiquitous yet enigmatic experience.

The intriguing connection between déjà vu and memory consolidation offers a fertile ground for investigation. Memory consolidation, the process by which the brain transforms fragile, short-term memories into stable, long-term ones, is a complex and multi-stage process involving various brain regions and neurochemical pathways. Disruptions or imperfections at any stage of this process could potentially contribute to the generation of déjà vu.

One potential mechanism lies in the encoding stage. Encoding refers to the initial process of transforming sensory information into a neural code that the brain can store. If the encoding process is incomplete or flawed, the resulting memory might be fragmented or distorted. When encountering a new situation that shares superficial similarities with this poorly encoded memory, the brain might experience a sense of familiarity, even though it lacks the detail or context to consciously recognize the source of that familiarity. This incomplete encoding could be due to various factors such as distractions, stress, or simply a lack of attention during the initial experience.

Imagine, for example, attending a large conference. You might briefly interact with several people, but due to the overall sensory overload, the encoding of these encounters might be incomplete. Later, encountering one of these individuals in a different setting might trigger a feeling of déjà vu, not because you fully remember the previous encounter but because a fragment of the incomplete memory is reactivated.

The storage phase of memory consolidation involves the stabilization and strengthening of newly encoded memories. During this stage, memories undergo structural and functional changes within the brain, including the strengthening of synaptic connections and the reorganization of neural networks. Errors during this phase could result in the mis-storage of memory traces, leading to distortions or confabulations. For instance, a memory might be mis-associated with a particular time or place, leading to the sensation that a currently experienced event has already occurred in a different context. This would be especially relevant for events that occurred during periods of sleep deprivation or intoxication, when memory consolidation is less efficient and prone to error. In such cases, the brain may incorrectly link a novel situation with a distorted, poorly stored memory, leading to the sensation of déjà vu. The subtle differences between the actual current situation and the recalled (although inaccurate) memory may not be detected consciously, resulting in the pervasive feeling of familiarity.

The retrieval stage of memory consolidation is equally crucial. Retrieval is the process of accessing stored memories. Errors during retrieval could involve misattribution of memory sources, leading to the mistaken belief that a current experience is a recollection of a past event. Imagine visiting a place reminiscent of a childhood home, yet you've never been there before. The similarities might activate a faint memory trace associated with a similar environment from your childhood, triggering a feeling of familiarity – a déjà vu experience. The brain may incorrectly tag this current experience with the vague recollection, resulting in a sense of familiarity without a clearly defined origin.

Furthermore, the interplay between different memory systems contributes to the complexity of déjà vu. Our memories are not stored in a monolithic fashion; rather, they are distributed across various brain regions specialized in processing different aspects of an experience, such as visual details, sounds, emotions, and context. Discrepancies or inconsistencies in the retrieval of information from these various memory systems could lead to a feeling of incongruity or a sense that something isn't quite right, which could contribute to the feeling of déjà vu. Consider a situation where the visual elements of a current scene strongly resemble a past experience, while other sensory or contextual details do not match. This mismatch between different aspects of the memory might create a sense of unease or a subtle feeling of strangeness, yet the predominant visual familiarity leads to a dominant sense of déjà vu.

The role of hippocampal function in memory consolidation is especially relevant. The hippocampus, a key structure in the brain's medial temporal lobe, plays a central role in forming new memories and integrating them into existing networks. Damage or dysfunction in the hippocampus has been linked to various memory disorders, including amnesia and difficulties in establishing new memories. Although a direct link between hippocampal dysfunction and déjà vu hasn't been definitively established, it's plausible that subtle disruptions in hippocampal function could contribute to errors in memory encoding, storage, or retrieval, thereby increasing the likelihood of experiencing déjà vu. It is conceivable that subtle variations in hippocampal activity, either in structural integrity or neurochemical modulation, could lead to inconsistent memory encoding and recall, potentially generating the peculiar sense of familiarity associated with déjà vu.

Moreover, the process of memory reconsolidation, where existing memories are reactivated and then restabilized, also plays a part. During this process, memories can be altered or updated, reflecting the integration of new information or the influence of current emotional states. This vulnerability to modification during reconsolidation opens the possibility for errors that might lead to the misattribution of familiarity. For example, a seemingly new situation might trigger a faint recollection, which then undergoes reconsolidation, incorporating elements of the new experience. The subsequent retrieval of this modified memory might result in a feeling of déjà vu, where the blended recollection feels both familiar and somehow different. The feeling of being 'slightly off' often associated with déjà vu may indeed result from this blend of old and newly added experiences.

Individual variations in memory consolidation efficiency could also explain the differences in the frequency of déjà vu experiences across individuals. Some individuals might have more efficient memory consolidation mechanisms, resulting in fewer errors during encoding, storage, and retrieval, making them less prone to déjà vu. In contrast, others might have less efficient memory consolidation processes, leading to a higher probability of errors and a greater likelihood of experiencing déjà vu. This efficiency could also be influenced by factors such as sleep quality, stress levels, and overall cognitive health.

Finally, the influence of neurotransmitters such as glutamate and acetylcholine, which play essential roles in memory consolidation, must not be overlooked. Fluctuations in the levels or activity of these neurotransmitters could affect the fidelity of memory encoding and retrieval, potentially leading to the generation of déjà vu. Further research could explore the interaction of these neurotransmitters with other factors, such as stress, sleep deprivation, and pre-existing neurological conditions, and their combined influence on the likelihood of experiencing déjà vu.

In conclusion, while the precise mechanisms underlying déjà vu remain elusive, the strong relationship between this phenomenon and the intricate process of memory consolidation is undeniable. Further research into the neurobiological underpinnings of memory, focusing on the encoding, storage, retrieval, and reconsolidation phases, promises to shed more light on this enigmatic experience, offering potential inroads into a deeper understanding of the human mind's remarkable, yet sometimes flawed, capacity for remembering. The fascinating aspect of déjà vu remains its enigmatic blend of familiarity and unfamiliarity, a testament to the malleability and complexity of human memory and the subtle ways in which our perceptions can shape our understanding of reality.

The neurobiological explanations for déjà vu, while compelling, only offer a partial understanding of this phenomenon. To fully grasp its significance, we must also consider its cultural and social dimensions. Déjà vu is not merely a neurological quirk; it's a shared human experience, woven into the fabric of our collective consciousness, interpreted and understood differently across various cultures and societies. Exploring these diverse interpretations provides valuable insights into how different societies perceive and interact with the enigmatic nature of memory, familiarity, and the subjective experience of time.

One of the striking aspects of déjà vu is its universality. Across diverse cultures, from the technologically advanced societies of the West to the more traditional communities of the East, individuals report experiencing déjà vu. This widespread occurrence suggests a fundamental aspect of human cognition, transcending cultural boundaries and societal norms. However, the *meaning* ascribed to déjà vu varies considerably. In some cultures, déjà vu is considered a relatively commonplace and unremarkable experience, a minor glitch in the system of memory, a fleeting sensation of familiarity quickly dismissed. In other cultures, déjà vu might carry more profound significance.

For example, in certain indigenous cultures, a déjà vu experience might be interpreted as a glimpse into the past, a visitation from ancestral spirits, or a sign from the divine. These interpretations often tie into a more holistic worldview, one that embraces a less rigid separation between the physical and spiritual realms. The feeling of familiarity might be understood not as a malfunction of the brain but as a meaningful connection to a broader, interconnected reality. This contrasts sharply with the more reductionist, materialistic approach of modern Western science, which primarily focuses on the neurobiological mechanisms underlying déjà vu, often neglecting or downplaying its potential spiritual or symbolic significance.

The prevalence of déjà vu within a particular culture can also be influenced by factors such as literacy rates, technological advancement, and the overall cultural emphasis on rationality versus intuition. Societies that emphasize scientific thinking and logical reasoning may tend to downplay or rationalize déjà vu, attributing it to simple cognitive processes. In contrast, cultures that place greater value on intuition and spiritual experiences might view déjà vu as more significant, potentially leading to a higher reported prevalence or a greater willingness to discuss such experiences openly. Anthropological studies comparing different cultures' responses to déjà vu could shed light on these subtle yet significant influences.

Furthermore, the way a culture frames memory itself can profoundly influence its interpretation of déjà vu. Cultures that emphasize oral traditions and the transmission of knowledge through storytelling might perceive déjà vu as a validation of past experiences, a resonance with memories transmitted through generations. In contrast, cultures with a strong emphasis on written records might consider déjà vu as an anomaly, a disruption of the orderly progression of time as meticulously documented.

The social implications of déjà vu are also worthy of consideration. The sharing of déjà vu experiences can foster a sense of shared reality and connection between individuals. Discussing a mutual feeling of familiarity, even in the absence of a shared memory, can create a sense of bonding and collective understanding. The act of verbalizing and validating the experience might itself influence its perception and significance. This social validation can reinforce the perceived reality and significance of the déjà vu, emphasizing its shared human quality. Conversely, the inability to share or explain a déjà vu experience can lead to feelings of isolation or even anxiety, highlighting the social dimension of the subjective experience.

The social context in which déjà vu occurs can also impact its perceived meaning. A déjà vu experience during a stressful situation might be interpreted differently than one occurring during a relaxed and contemplative moment. The emotional context coloring the experience might shape how the event is remembered and reported. These contextual factors need to be accounted for in any comprehensive analysis of cultural responses to déjà vu.

Research into the sociocultural aspects of déjà vu is still in its early stages. Comparative studies across various cultures, employing both quantitative and qualitative methodologies, are needed to fully explore the cultural nuances of this phenomenon. Such studies might investigate differences in the frequency of reported déjà vu experiences, the types of interpretations given to these experiences, and the impact of social and cultural contexts on the perception and significance of déjà vu. Understanding the cultural context within which déjà vu is framed will give us a more comprehensive understanding of how it is experienced and understood by people around the world.

The investigation into the cultural and social aspects of déjà vu could also offer valuable insights into other areas of cognitive science and psychology. It could illuminate the relationship between personal memory, collective memory, and cultural narratives. It could also shed light on the ways in which societies construct and maintain shared realities, and how cultural beliefs influence the interpretation and meaning-making processes related to personal experiences. This understanding, in turn, could provide important insights for improving communication and understanding across cultures, particularly in areas where cultural differences might lead to misinterpretations and misunderstandings.

The seemingly simple phenomenon of déjà vu, therefore, opens up a world of inquiry into the fascinating interplay between neurobiology, culture, and subjective experience. Its universality highlights a fundamental aspect of human cognition, while its diverse interpretations across cultures underscore the crucial role of context and societal beliefs in shaping our understanding of reality. Further exploration of this phenomenon, through interdisciplinary research combining neuroscience, anthropology, psychology, and sociology, holds the key to unlocking a more profound understanding of the human experience itself – its mysteries, its shared realities, and its rich tapestry of diverse perspectives. The study of déjà vu thus transcends the mere investigation of a neurological oddity; it becomes a window into the depths of the human mind and the multifaceted ways in which we interact with, interpret, and create our own reality.

Having explored the neurobiological and sociocultural facets of déjà vu, we now turn to a more speculative, yet potentially illuminating, avenue of inquiry: its possible connection to non-linear time theories. While firmly rooted in the realm of theoretical physics and metaphysics, these theories offer intriguing frameworks through which to interpret the seemingly paradoxical nature of déjà vu. The feeling of having lived a moment before, even when rationally knowing it to be impossible, challenges our conventional linear understanding of time as a unidirectional, ever-progressing flow.

Non-linear time theories, such as those found in some interpretations of quantum mechanics and certain cosmological models, posit that time may not be the singular, linear progression we perceive it to be. Instead, they suggest the existence of multiple timelines, parallel universes, or a more complex, multi-dimensional temporal structure. Within these frameworks, déjà vu might be interpreted not as a malfunction of memory, but as a fleeting glimpse into a parallel reality, a momentary bleed-through from another timeline or dimension.

This interpretation gains credence when considering the subjective nature of déjà vu. The experience is intensely personal and often lacks any readily identifiable trigger or corresponding memory. The sensation of familiarity is not tied to a specific past event, but rather to a feeling of *having been* in that very moment before, a feeling that defies conventional temporal logic. This elusive quality aligns remarkably well with the notion of accessing or briefly experiencing a parallel reality, one where the exact same scene, conversation, or sensory experience may have already unfolded.

Furthermore, some proponents of non-linear time theories argue that our perception of time is fundamentally limited, that our consciousness is only capable of perceiving a small sliver of the totality of temporal reality. Déjà vu, in this context, could represent a momentary breach of this perceptual barrier, a spontaneous glimpse into a hidden dimension of time, a reality slightly offset from our own. This suggests that our brains might possess, however subtly, the capacity to detect or interact with these other temporal streams, resulting in the fleeting, yet intensely powerful, sensation of familiarity.

One intriguing aspect of this perspective is the potential explanation it offers for the seemingly random nature of déjà vu experiences. If they represent glimpses into parallel realities, their unpredictable occurrences would be consistent with the inherent uncertainty and probabilistic nature often associated with quantum phenomena. Just as quantum entanglement involves instantaneous correlations between seemingly disparate particles, regardless of spatial separation, so too might déjà vu reflect an instantaneous connection or interaction between parallel timelines.

The notion of parallel universes resonates with some interpretations of quantum mechanics, particularly the Many-Worlds Interpretation (MWI). MWI proposes that every quantum measurement causes the universe to split into multiple branches, each representing a different possible outcome. In this framework, every decision we make, every event that unfolds, generates countless alternative universes, each following a distinct trajectory. Déjà vu, within the MWI framework, could represent a brief, unintended interaction with one of these parallel universes, a momentary awareness of a reality where a particular event has already transpired.

However, it is crucial to acknowledge the profound speculative nature of this interpretation. While non-linear time theories are subject to ongoing scientific debate and investigation, the connection between these theories and déjà vu remains firmly within the realm of hypothesis and conjecture. Current scientific methods lack the capacity to directly test or verify the existence of parallel universes or other non-linear temporal phenomena, making any definitive conclusions regarding their relationship to déjà vu currently impossible.

Nevertheless, exploring such theoretical possibilities can be valuable. By engaging with seemingly outlandish ideas, we can broaden our understanding of the limitations of our current scientific models and push the boundaries of our understanding of consciousness and the nature of reality itself. Even if the non-linear time interpretation of déjà vu ultimately proves untenable, the very act of exploring this possibility challenges our preconceived notions about time, memory, and the nature of experience.

It's also important to consider the philosophical implications of this approach. If déjà vu is indeed linked to non-linear time, it raises profound questions about the nature of identity, free will, and the perceived continuity of self. If we can momentarily glimpse other realities where our lives unfolded differently, does this challenge the notion of a fixed, singular self? Does it suggest a more fluid, multi-dimensional conception of selfhood?

The investigation into the potential link between déjà vu and non-linear time theories underscores the intricate interplay between science and philosophy, the constant dance between empirical observation and theoretical speculation. It highlights the limitations of current scientific paradigms and the need for interdisciplinary approaches to unravel the mysteries of human consciousness and the seemingly paradoxical nature of our experiences.

Moreover, the consideration of non-linear time perspectives extends beyond the realm of pure physics and opens doors to broader discussions within cognitive science and consciousness studies. Some researchers propose that our subjective experience of time is not merely a passive reflection of external events but an active construction of our minds. This perspective suggests that the brain may be actively shaping and interpreting our experience of time in ways that are not fully understood, potentially allowing for glimpses into other temporal possibilities.

The exploration of such perspectives compels us to confront the fundamental limitations of our current scientific understanding of time and consciousness. We are forced to question the assumptions that underpin our linear perception of time and to grapple with the possibility that our reality is far richer and more complex than we currently perceive it to be.

In conclusion, while a direct causal link between déjà vu and non-linear time theories remains largely speculative, exploring this potential connection prompts a significant re-evaluation of our understanding of time, memory, and consciousness. It encourages a more holistic and interdisciplinary approach to the study of déjà vu, one that integrates insights from neuroscience, physics, philosophy, and psychology. While we may not yet possess the scientific tools to definitively prove or disprove these theories, the very act of considering them expands the boundaries of our inquiry, opening up new avenues of exploration that might ultimately lead to a deeper comprehension of this enigmatic phenomenon and the nature of reality itself. The seemingly simple experience of déjà vu, therefore, becomes a gateway to exploring profound questions about the nature of time, consciousness, and the very structure of our reality.

13 Exploring the Physics of Time

Our journey into the potential physics behind déjà vu now leads us to arguably the most profound revolution in our understanding of time: Albert Einstein's theories of relativity. Before Einstein, time was considered an absolute, a universal constant ticking away uniformly for everyone, everywhere. Newtonian physics, the dominant paradigm for centuries, treated time as a backdrop against which the events of the universe played out. But Einstein shattered this seemingly self-evident notion, revealing a far more intricate and dynamic relationship between space and time.

Einstein's special theory of relativity, published in 1905, fundamentally altered our understanding of simultaneity. The core postulate of special relativity is that the laws of physics are the same for all observers in uniform motion, and the speed of light in a vacuum is the same for all observers, regardless of the motion of the light source. This seemingly simple statement has profound consequences. Consider two observers, one stationary and the other moving at a high velocity relative to the first. If a light signal is emitted from a point midway between them, Newtonian physics would suggest both observers would see the light signal arrive simultaneously. However, special relativity dictates otherwise. Due to the constancy of the speed of light, the moving observer will perceive the light signal reaching them before the stationary observer, even though the light signal traveled the same distance for both. This means that simultaneity itself is relative; what is simultaneous for one observer is not necessarily simultaneous for another.

This relativity of simultaneity is intricately linked to time dilation, another crucial consequence of special relativity. Time dilation refers to the phenomenon where time appears to slow down for an observer moving at a high velocity relative to a stationary observer. The faster an object moves, the slower time passes for that object relative to a stationary observer. This is not a subjective illusion; it's a real physical effect that has been experimentally verified countless times. For example, highly accurate atomic clocks flown on high-speed aircraft have shown a measurable time difference compared to identical clocks left on the ground.

The effect, while minuscule at everyday speeds, becomes significant at speeds approaching the speed of light. If a spacecraft were to travel at a substantial fraction of the speed of light, the time experienced by the astronauts onboard would be slower than the time experienced by observers on Earth. Upon their return, the astronauts would find that less time had passed for them than for their counterparts who remained on Earth. This seemingly paradoxical effect is a direct consequence of the constancy of the speed of light and the interwoven nature of space and time.

Einstein's general theory of relativity, published in 1915, expanded upon special relativity by incorporating gravity. General relativity describes gravity not as a force, as Newton had proposed, but as a curvature of spacetime caused by the presence of mass and energy. Imagine a bowling ball placed on a stretched rubber sheet; the ball creates a dip, causing other objects rolling on the sheet to curve towards it. Similarly, massive objects warp spacetime, creating gravitational fields that influence the motion of other objects.

This curvature of spacetime has profound implications for our understanding of time. Time is not only relative to velocity, as in special relativity, but it is also affected by the strength of the gravitational field. In regions of strong gravitational fields, time slows down relative to regions of weaker gravitational fields. This effect, known as gravitational time dilation, has been experimentally verified by comparing atomic clocks at different altitudes. Clocks placed at higher altitudes, where the gravitational field is slightly weaker, run slightly faster than clocks placed at lower altitudes.

The implications of general relativity for our perception of time are far-reaching. The universe is not a static, unchanging entity, but a dynamic, evolving system where time itself is influenced by the distribution of mass and energy. The expansion of the universe, for instance, affects the flow of time on cosmological scales. As the universe expands, the distances between galaxies increase, leading to changes in the way time unfolds across vast cosmic distances.

Furthermore, the general theory of relativity has profound implications for the nature of black holes, regions of spacetime with such strong gravitational fields that nothing, not even light, can escape. The extreme curvature of spacetime near a black hole causes time to slow down dramatically, approaching a complete standstill at the black hole's singularity.

For an observer watching an object falling into a black hole, the object would appear to slow down and fade away, never actually crossing the event horizon. This illustrates the extreme nature of gravitational time dilation and the interconnectedness of time and space.

The connection between Einstein's theories of relativity and our subjective experience of time, such as déjà vu, is far from straightforward. Relativistic effects, while real and measurable, are generally too small to account for the subjective experience of having already lived a moment. However, the conceptual shift brought about by relativity – the understanding that time is not absolute but relative to the observer's frame of reference – opens up avenues for speculative thought.

One could argue that the non-linearity hinted at in some interpretations of quantum mechanics, as discussed earlier, might find a tenuous but intriguing connection to relativity. The highly curved spacetime near black holes, for example, presents a scenario where our linear understanding of time might break down. While not a direct explanation for déjà vu, it's a reminder that our perception of time, as linear and unidirectional, might be a limited interpretation of a far more complex reality.

Moreover, Einstein's work challenges the very foundations of our intuitive understanding of time. The notion that time is relative, interwoven with space, and affected by gravity suggests a far more dynamic and flexible understanding of temporal reality than the simple linear progression we typically perceive. The potential existence of multiple universes, as suggested by some interpretations of quantum mechanics, could even be seen as a kind of "super-relativity," where different temporal realities coexist, though the mechanisms connecting them remain purely speculative.

In conclusion, Einstein's theories of relativity, though not providing a direct explanation for déjà vu, profoundly impact our understanding of time's nature. The relativity of simultaneity and time dilation, along with the curvature of spacetime described by general relativity, show that time is not a universal constant but a dynamic quantity, deeply intertwined with the fabric of the universe itself. While the connection to déjà vu remains speculative, the profound implications of relativity challenge our preconceived notions of temporal linearity and open new avenues of exploring the mysterious nature of time and subjective experience. The very act of considering these ideas, even in the context of seemingly inexplicable phenomena like déjà vu, pushes the boundaries of our understanding, paving the way for future insights into the mysteries of consciousness and the cosmos. The investigation continues, weaving together threads of physics, philosophy, and potentially, the enigmatic world of parapsychology, in an attempt to unravel the secrets of time and its impact on the human experience.

Having explored the relativistic revolution in our understanding of time, we now turn our attention to another fundamental aspect of temporal experience: the arrow of time. Unlike the relativity of simultaneity and time dilation, which are consequences of the structure of spacetime, the arrow of time points to a deeper mystery—the seemingly irreversible flow of time from past to future. This unidirectional flow is a ubiquitous feature of our everyday experience. We remember the past, anticipate the future, but never experience them in reverse order. A dropped glass shatters, but we never observe the shards spontaneously reassembling themselves into a whole glass. This seemingly simple observation, however, has profound implications for our understanding of the universe.

Classical mechanics, the physics of Newton, is time-symmetric. Its equations work equally well whether time moves forward or backward. If you reversed the velocities of all the particles in a Newtonian system, the system would retrace its steps perfectly, demonstrating a perfect time-reversal symmetry. This time symmetry is inherent in the fundamental equations of classical mechanics, suggesting that the past and future are equally accessible. But our lived experience contradicts this. The arrow of time, pointing inexorably towards the future, appears to be a fundamental asymmetry in the universe, one that classical physics struggles to explain.

The answer, it seems, lies in the realm of thermodynamics, specifically in the concept of entropy. Entropy, a measure of disorder or randomness in a system, is fundamentally linked to the arrow of time. The second law of thermodynamics states that the total entropy of an isolated system can only increase over time or remain constant in ideal cases where the system is in a steady state or undergoing a reversible process. This law is statistical in nature, meaning it describes the overwhelming probability of events unfolding in a specific direction. It does not forbid a decrease in entropy; it simply states that it is exceedingly unlikely.

Imagine a deck of cards perfectly ordered by suit and rank. If you shuffle the deck, the entropy increases, resulting in a random arrangement. While it is theoretically possible to shuffle the deck back to its original order, the probability of this happening randomly is astronomically low. The second law of thermodynamics reflects this inherent bias towards increasing disorder. The spontaneous transition from order to disorder is far more probable than the reverse transition.

The connection between entropy and the arrow of time is profound. The increase in entropy provides a physical basis for the unidirectional flow of time that we experience. The past is characterized by lower entropy (greater order) while the future is characterized by higher entropy (greater disorder). This increasing entropy is not a fundamental law of physics like the conservation of energy, but a statistical tendency arising from the vast number of possible microstates that a system can occupy.

However, the explanation of the arrow of time based on increasing entropy is not without its challenges. The universe began in a state of extraordinarily low entropy—the Big Bang—a highly ordered state from which the universe has expanded and evolved into a state of greater disorder. Why the universe started in such a low-entropy state remains a profound and open question, often referred to as the "entropy problem." Several theories attempt to address this, including the idea of multiple universes or the possibility that the universe's apparent low initial entropy is a result of an underlying, yet unknown, physical principle.

Another challenge arises from the seeming contradiction between the time-symmetric nature of fundamental physical laws (excluding the second law of thermodynamics) and the time-asymmetric nature of our everyday experience. The fundamental equations of both classical and quantum mechanics are largely time-reversible, yet the world around us is governed by an irreversible arrow of time. This discrepancy hints at a deeper mystery, possibly indicating a profound gap in our understanding of the very fabric of reality.

Beyond the standard thermodynamic explanation, the nature of time itself, its linearity, and whether this apparent linearity is accurate are questions physicists and philosophers alike constantly revisit. While our everyday experience overwhelmingly supports a linear perception of time, with events proceeding in a clearly defined sequence from past to future, certain physical scenarios suggest the possibility of non-linear time, or even time's complete absence.

The concept of non-linear time challenges our intuitive understanding of causality. In a linear timeline, cause precedes effect; events unfold in a sequential manner. However, certain interpretations of quantum mechanics suggest that cause and effect might not be strictly sequential, leading to the potential for non-linear temporal relationships. The famous double-slit experiment, for instance, demonstrates that the act of observation can influence the outcome of an event, suggesting a breakdown of strict causal order in the quantum realm. While these quantum effects may not directly impact our macroscopic world in observable ways, they highlight the limitations of our classical understanding of causality and the potential for more complex temporal relationships.

Moreover, theoretical concepts like wormholes, predicted by general relativity, offer the possibility of traversing time in non-linear ways. While the feasibility of traversing wormholes remains highly speculative, their theoretical existence opens up the possibility of shortcuts through spacetime, potentially linking different points in time. These theoretical constructs, however, are often accompanied by paradoxes, such as the grandfather paradox, which arises from the potential of altering past events and their cascading consequences on the present. Resolving these paradoxes requires carefully considering the nature of time, causality, and the very possibility of altering the past.

The exploration of non-linear time leads us to confront deeper philosophical questions about the nature of reality. If time is not linear, are past, present, and future coexistent in some manner? Does free will exist in a universe where events might be interconnected in non-sequential ways? These questions lie at the intersection of physics, philosophy, and potentially, even consciousness studies. The quest to understand the arrow of time is intrinsically linked to the ongoing search for a comprehensive theory of everything, a unified framework that can reconcile the seemingly contradictory aspects of our physical reality, including the apparent dichotomy between the reversible laws of physics and the irreversible flow of time.

The investigation into the arrow of time and its relation to entropy, thermodynamics, and the fundamental laws of physics continues. This exploration pushes the boundaries of our current understanding, unveiling deeper questions about the nature of time, causality, and the universe itself. The implications are vast, touching on the very core of our perception of reality and our place within the cosmos. As we delve deeper into the mysteries of time, we discover that the seemingly simple experience of a unidirectional flow masks an astonishing complexity, inviting us to explore the limits of our understanding and venture into the unknown frontiers of temporal reality. The arrow of time, it seems, points not just to the future, but to a deeper comprehension of the universe itself. The journey is far from over; the mysteries of time continue to beckon.

The second law of thermodynamics, a cornerstone of physics, provides a compelling, albeit incomplete, explanation for the arrow of time. This law, simply stated, dictates that the total entropy of an isolated system will either increase over time or, in ideal, reversible processes, remain constant. Entropy, often described as a measure of disorder or randomness, plays a crucial role in understanding the universe's seemingly irreversible progression. A system's entropy increases as it transitions from a state of order to disorder. Think of a neatly stacked deck of cards. The initial state exhibits low entropy – high order. Shuffling the deck introduces disorder, increasing its entropy. While theoretically possible, the spontaneous re-ordering of the shuffled deck is astronomically improbable. This inherent bias towards disorder is the essence of the second law.

The link between entropy increase and the arrow of time lies in the statistical likelihood of events. The universe, at its most fundamental level, is governed by probabilistic laws. The second law doesn't prohibit entropy decrease; it merely states that such a decrease is overwhelmingly improbable, given the vast number of possible microstates a system can occupy. The past, therefore, is statistically characterized by lower entropy (higher order), while the future is associated with higher entropy (higher disorder). This statistical bias, this overwhelming probability of proceeding from order to disorder, gives us the perceived unidirectional flow of time. We experience the universe evolving towards states of increased entropy, from the breaking of a dropped glass to the cooling of a hot cup of coffee.

However, this thermodynamic explanation of the arrow of time presents its own challenges. The universe itself began in a state of exceptionally low entropy – the Big Bang. This initial condition, a state of extreme order, presents a significant puzzle. If the universe is constantly evolving towards states of greater disorder, then why did it start in such an ordered state? This question, often referred to as the "entropy problem," remains one of the most significant unresolved issues in cosmology. Several hypotheses attempt to address this puzzle, including proposals involving multiverse theories, where our universe's low initial entropy might simply be a statistically improbable but not impossible occurrence amongst a vast ensemble of universes, each with varying initial conditions. Other theories posit the existence of an underlying physical principle, currently unknown to us, that explains the universe's initial low-entropy state.

Moreover, the thermodynamic explanation isn't entirely satisfactory because it doesn't address the apparent contradiction between the time-reversible nature of fundamental physical laws and the time-irreversible nature of our experiences. The fundamental equations of both classical and quantum mechanics are, to a large extent, time-symmetric. This means they remain valid if the direction of time is reversed. Reversing the velocities of all particles in a classical system would, theoretically, retrace the system's path perfectly. This time-symmetry is not observed in the macroscopic world, where the arrow of time is undeniably present. This discrepancy highlights a gap in our current understanding. It suggests that the macroscopic irreversibility we observe may emerge from the microscopic reversibility of the underlying physical laws, a transition that demands further investigation.

The statistical nature of the second law also leaves room for questions. While the probability of a decrease in entropy is exceptionally low, it is not strictly zero. Fluctuations in entropy are possible, albeit exceedingly rare. On a microscopic scale, we might observe localized decreases in entropy, but these are quickly overwhelmed by the overall increase in entropy within a larger system. This points to the crucial role of scale in understanding entropy and its relation to the arrow of time. The macroscopic irreversibility is a consequence of the statistics of a vast number of microscopic particles interacting probabilistically.

Beyond the standard thermodynamic approach, several other perspectives offer insights into the arrow of time. The role of information, for instance, has gained prominence. Some physicists argue that the arrow of time is deeply connected to the increase of information in the universe. As systems evolve, they generate information, creating more complex structures and patterns. This perspective suggests that the flow of time is tied to the ongoing creation and processing of information. This view connects the seemingly abstract notion of entropy with the concrete reality of information generation, suggesting that the universe's evolution is, in a sense, an ongoing process of information creation.

Furthermore, exploring the nature of time's linearity itself adds another layer of complexity. Our everyday experience profoundly supports a linear understanding of time. Events unfold sequentially, with a clear distinction between past, present, and future. However, some interpretations of quantum mechanics hint at a more nuanced picture. The superposition principle, which allows quantum systems to exist in multiple states simultaneously, challenges the strict linearity of causality. The famous double-slit experiment beautifully illustrates this: the very act of observation seemingly affects the outcome of the experiment, blurring the lines of strict causal order. This, however, doesn't necessarily imply a violation of causality but rather a limitation of our classical understanding of cause and effect within the quantum realm.

The theoretical concept of wormholes, predicted by Einstein's general theory of relativity, further complicates our understanding of time's linearity. Wormholes, hypothetical tunnels through spacetime, could potentially allow for non-linear time travel. While traversing a wormhole remains firmly in the realm of science fiction at present, the theoretical possibility opens avenues for questioning the very fabric of spacetime and the concept of linearity in temporal progression. However, such scenarios inevitably lead to paradoxes, such as the famous grandfather paradox, which explores the potential causal contradictions arising from altering the past. These paradoxes underscore the complex relationship between time, causality, and the consistency of our universe.

In summary, the arrow of time is a profound mystery that touches upon the very core of our understanding of the universe. While the second law of thermodynamics offers a partial explanation through the concept of increasing entropy, several unresolved questions remain. The universe's low initial entropy, the discrepancy between the time-reversible nature of fundamental laws and the time-irreversible nature of our experience, and the potential non-linearity of time itself all point to a deeper understanding that is yet to be reached. The ongoing exploration of these questions necessitates a multidisciplinary approach, combining insights from physics, thermodynamics, philosophy, and even potentially cognitive science and parapsychology, to fully comprehend the nature of time and its unidirectional flow. The journey toward a complete understanding remains open, revealing the universe's complex tapestry of temporal dynamics, one thread at a time. The pursuit continues, a testament to humanity's enduring curiosity about the fundamental nature of reality. The arrow of time, it seems, points not just toward a future yet unknown but also towards a deeper comprehension of the universe itself.

Our exploration of time's arrow has thus far focused primarily on thermodynamic considerations, specifically the second law and its implications for entropy. However, a complete understanding requires venturing into the realm of quantum gravity, a field striving to reconcile the seemingly irreconcilable: Einstein's general theory of relativity, which describes gravity and the large-scale structure of the universe, and quantum mechanics, which governs the behavior of matter at the subatomic level. The incompatibility between these two highly successful theories remains one of the most significant challenges in modern physics, and its resolution is intimately linked to our comprehension of time.

General relativity portrays time as inextricably intertwined with space, forming the four-dimensional fabric of spacetime. Gravity, according to Einstein, isn't a force but rather a manifestation of the curvature of spacetime caused by the presence of mass and energy. This curvature dictates the paths of objects moving through spacetime, including the passage of time itself. Time's flow, in this framework, isn't uniform; it's relative, affected by gravity's strength. Time slows down near massive objects, a phenomenon confirmed by numerous experiments, most famously with atomic clocks at differing altitudes. This relativistic effect suggests time's fluidity, its dependence on the gravitational field.

Quantum mechanics, on the other hand, operates on a vastly different scale, dealing with the probabilistic nature of subatomic particles. Time, within the quantum mechanical framework, is typically treated as a background parameter, an external variable against which quantum events unfold. The theory doesn't inherently address the nature of time itself but rather utilizes it as a coordinate for describing the evolution of quantum states. This stark contrast between the continuous, deterministic spacetime of general relativity and the probabilistic, discrete nature of quantum mechanics creates a fundamental tension.

The search for a theory of quantum gravity aims to bridge this chasm, uniting the macroscopic picture of spacetime with the microscopic realm of quantum phenomena. Several promising approaches exist, each attempting to grapple with the fundamental questions about time's nature within this unified framework. Loop quantum gravity, for instance, proposes that spacetime itself is quantized, composed of fundamental units of space and time, analogous to the quantization of energy in quantum mechanics. In this model, time might emerge as a property of the underlying quantum structure of spacetime, rather than being a fundamental parameter. This perspective potentially addresses the issue of time's arrow by suggesting that the directionality of time could arise from the dynamics of these fundamental quantum structures.

String theory, another leading candidate for a quantum gravity theory, presents an even more radical departure from our classical understanding of spacetime. It postulates that fundamental constituents of the universe aren't point-like particles but rather tiny, vibrating strings. The different vibrational modes of these strings correspond to different particles and forces. In string theory, spacetime emerges as a consequence of the interactions between these strings, suggesting that time's very existence is a derived property. This perspective challenges the notion of time as a pre-existing backdrop, instead suggesting it as an emergent phenomenon, intricately linked to the quantum dynamics of these fundamental strings.

These theoretical frameworks offer tantalizing possibilities for addressing the enigma of time's arrow within a fully quantized theory of gravity. If spacetime itself is quantized, as proposed by loop quantum gravity, then the statistical mechanics governing the behavior of these fundamental units might intrinsically incorporate an arrow of time. The transition from a low-entropy state to a high-entropy state could be a fundamental consequence of the quantum dynamics of spacetime, resolving the puzzle of the universe's initial low entropy and the seeming contradiction between time-reversible microscopic laws and time-irreversible macroscopic observations.

Similarly, in string theory, the emergence of spacetime from the underlying string dynamics could provide a natural explanation for the arrow of time. The process by which spacetime emerges might inherently incorporate a directionality, leading to the observed unidirectional flow of time. The complexity of these theories, however, makes it incredibly difficult to extract definitive predictions regarding the arrow of time. The mathematical intricacies of both loop quantum gravity and string theory are formidable, and the extraction of experimentally testable predictions remains a major challenge.

Beyond these major theoretical frameworks, other approaches to quantum gravity are being explored, including causal set theory, which suggests that spacetime is fundamentally discrete, composed of a partially ordered set of events. This perspective could provide a novel way to incorporate the arrow of time into a quantum gravity framework. The idea that time isn't continuous but rather a series of discrete events could help explain the irreversibility we observe in the macroscopic world.

The challenges in experimentally testing these theories are immense. We lack the technological capability to directly probe the quantum structure of spacetime. The energies required to explore these scales are far beyond our current reach. Indirect tests, however, are being sought through various avenues, including the study of black holes, where the effects of strong gravity and quantum phenomena intertwine. Observations of Hawking radiation, for instance, could provide insights into the quantum nature of spacetime near a black hole's event horizon, potentially offering clues to the nature of time in these extreme environments.

The implications of a unified theory of quantum gravity for our understanding of time extend far beyond theoretical physics. Such a theory could fundamentally alter our philosophical perspectives on the nature of reality, causality, and free will. If time is an emergent property, as suggested by some approaches, then the very concept of a linear, unidirectional flow of time might need revision. The implications for our understanding of consciousness and the subjective experience of time are profound, potentially opening doors to exploring the connections between physics and consciousness in unprecedented ways.

The quest to unify general relativity and quantum mechanics, therefore, isn't merely a pursuit of theoretical elegance; it's a journey into the very heart of our understanding of reality. The resolution of this profound scientific challenge holds the key to unlocking some of the most fundamental mysteries of the universe, including the enduring enigma of time's arrow. The development of a successful theory of quantum gravity promises not only to explain the origins of time's unidirectional flow but also to shed light on the deepest aspects of the cosmos and our place within it. The endeavor continues, pushing the boundaries of human knowledge and revealing, step by step, the intricate tapestry of time's mysteries. The future of this research remains as open and potentially revolutionary as the very nature of time itself.

The conceptual leap of treating time as a dimension, analogous to the three spatial dimensions, is profound. In classical Newtonian physics, time marches relentlessly forward, a universal backdrop against which events unfold. However, Einstein's theory of special relativity revolutionized this view. By demonstrating the equivalence of space and time, and their intertwining to form spacetime, relativity revealed time's malleability. It's no longer an absolute, invariant quantity but rather a relative one, dependent on the observer's velocity and the gravitational field. This relativity of simultaneity means that events considered simultaneous in one frame of reference might not be so in another, shattering the classical notion of a universally agreed-upon "now."

The implications of spacetime are far-reaching. Consider the phenomenon of time dilation. As an object approaches the speed of light, time slows down for that object relative to a stationary observer. This isn't just a theoretical construct; it's been experimentally verified with atomic clocks flown on high-speed jets and orbiting satellites. The Global Positioning

System (GPS), a technology we rely on daily, requires precise timekeeping, and it wouldn't function accurately without accounting for relativistic time dilation effects. The clocks on GPS satellites, experiencing weaker gravitational fields than those on Earth, tick slightly faster than ground-based clocks. These minute differences, if unaccounted for, would lead to significant navigational errors over time.

General relativity extends this concept further, incorporating gravity into the fabric of spacetime. Mass and energy warp spacetime, causing objects to follow curved paths, as described by the geodesic principle. Time itself is affected by this curvature; it slows down in stronger gravitational fields. This is known as gravitational time dilation. A clock placed near a massive object, like a black hole, would tick slower than an identical clock far away from the gravitational influence. This effect is not merely a subtle curiosity; it has significant astrophysical consequences, affecting the timing of events in strong gravitational fields. The precise measurement of gravitational time dilation near compact objects offers a crucial test of general relativity.

The treatment of time as a dimension opens up possibilities beyond the confines of relativistic physics. Imagine traversing time, not merely observing its passage. While currently relegated to science fiction, the concept of time travel raises profound questions about causality and the nature of reality. Could we, hypothetically, move backward or forward in time, altering past events or influencing the future? The paradoxes associated with time travel, such as the "grandfather paradox," highlight the inherent difficulties in reconciling such concepts with our intuitive understanding of cause and effect. The resolution of these paradoxes likely requires a deeper understanding of the fundamental nature of time and its relationship to other physical laws.

Furthermore, considering time as a dimension allows us to approach the problem of the universe's origin and evolution from a fresh perspective. The Big Bang theory posits a singularity, a point of infinite density and temperature from which the universe began. But what does this mean for time? Did time begin at the Big Bang, or was it a pre-existing entity? Cosmology grapples with this question, attempting to describe the universe's evolution as a trajectory through spacetime. Models of the universe's evolution, from its early stages to its eventual fate, rely on mathematical descriptions of spacetime, emphasizing time's role as an integral dimension.

Beyond the purely physical realm, the dimensional nature of time has intriguing implications for our understanding of consciousness and subjective experience. Our perception of time's flow—its seeming unidirectionality—is a fundamental aspect of human consciousness. We remember the past, experience the present, and anticipate the future. Yet, the physical laws governing the universe are, at a fundamental level, largely time-reversible. This disparity between our subjective experience of time and the time-symmetric nature of many physical laws poses a fascinating challenge for both physics and philosophy. Many theories attempt to bridge this gap. Some posit that the arrow of time is an emergent property arising from the statistical behavior of many particles.

However, exploring the potential for non-linearity in time throws a wrench into this traditional understanding. If time isn't simply a linear progression, but instead a multidimensional construct, a complex network of interconnected moments, our perceptions could be dramatically altered. The very notion of causality might change, suggesting a universe less predictable, more entangled, and potentially less constrained by linear cause and effect. This possibility opens up avenues of investigation into the very nature of free will. If time is non-linear, the supposed determinism of the universe might be challenged, allowing room for genuine choice and agency in a manner that linear time struggles to accommodate.

The exploration of non-linear time requires a deeper delve into advanced theoretical frameworks. The concept of spacetime itself, as presented in general relativity, already introduces a degree of non-linearity through its curved geometry. But going beyond this, consider theories that propose multiple time dimensions, or that the fabric of time itself might fold back on itself, creating temporal loops or closed timelike curves. These ideas, while speculative, challenge our fundamental notions of time and open up possibilities that currently lie beyond our ability to empirically verify.

Furthermore, some interpretations of quantum mechanics introduce a degree of non-linearity into the temporal realm. The uncertainty principle, for example, imposes inherent limitations on the simultaneous knowledge of a particle's position and momentum, effectively blurring the sharp distinctions that define classical, deterministic time. Similarly, the concept of quantum entanglement, where two or more particles become interconnected regardless of the spatial separation, potentially implies a non-local connection between events in time, challenging the strict linear progression of cause and effect.

The quest to understand time's nature, therefore, remains an open frontier of scientific inquiry. From the relativistic warping of spacetime to the quantum fuzziness of time's flow, and the possibility of non-linear temporal structures, the mysteries surrounding time continue to challenge and inspire. The implications of our exploration extend far beyond the realm of theoretical physics; they profoundly impact our understanding of the universe's origin, evolution, and even the nature of consciousness and reality itself. The journey to unravel time's secrets is a journey into the very heart of existence, a journey that promises to reveal unforeseen depths and complexities as we progress. The future of this understanding remains unwritten, as open and potentially revolutionary as time itself.

14 Embracing Non-Linearity

In conclusion, film and television offer a remarkably versatile canvas for representing non-linear time. From the simple use of flashbacks and flashforwards to the more complex manipulation of narrative structure, visual elements, and subjective experience, filmmakers have developed a range of techniques to explore the multifaceted nature of temporality. By skillfully manipulating these tools, filmmakers can create compelling narratives, explore complex themes, and engage viewers in a profound way, enriching our understanding of the fluid and often unpredictable nature of time itself. The ability to manipulate time in film is not just a technical tool but a creative force, fundamentally shaping our engagement with stories and pushing the boundaries of cinematic expression. The dynamic interplay between technical skill and artistic vision underscores the profound power of cinematic representation in shaping our perception of time, moving beyond mere linear chronology to embrace the complexities and paradoxes of temporal experience.

Our ingrained perception of time as a linear, unidirectional flow—a river relentlessly moving from past to future—is a deeply ingrained aspect of human consciousness. This linear model, while useful for navigating daily life, is demonstrably inadequate when confronted with the complexities revealed by modern physics and emerging perspectives in cognitive science. The limitations of this linear view become especially apparent when considering phenomena that challenge the very foundations of causality and temporal sequence.

Consider, for instance, the subjective experience of time. For many, time seems to accelerate as we age; childhood summers stretching into an eternity, while the years of adulthood whiz by in a blur. This subjective distortion of temporal experience cannot be explained solely by the ticking of a clock. The emotional intensity associated with certain memories, the level of engagement in an activity, and even our physiological state all contribute to our individual perception of temporal passage. This inherent subjectivity underscores the inadequacy of a purely objective, linear model of time.

Furthermore, the study of memory reveals further complexities. Our memories are not simply recordings of past events; they are actively reconstructed each time we access them. The act of remembering involves a process of interpretation and re-integration, influencing the very structure and content of our recollections. This implies a dynamic, interactive relationship between past, present, and future, where memories are not static entities but rather fluid constructs subject to ongoing modification and influence.

The apparent fluidity of memory challenges the linear model's rigid structure. If our memories are subject to change, how can we be certain of the accuracy or stability of our perceptions of the past? This blurring of temporal boundaries suggests a more interconnected, potentially even non-linear, relationship between past, present, and future. Events that occurred seemingly long ago can exert influence on our present lives, shaping our decisions and behaviors in ways that a strictly linear model cannot fully explain.

The concept of precognition, while controversial, further illustrates the inadequacy of a linear temporal framework. If individuals can accurately predict future events, then the future must, in some sense, exert influence on the present, violating the strict causal sequence of the linear model. While many explanations for apparent precognitive abilities focus on chance, coincidence, or subconscious processing, the persistence of such reports merits serious consideration, especially in the context of our evolving understanding of time.

Similarly, the widespread phenomenon of déjà vu, that uncanny feeling of having already experienced the present moment, defies a purely linear interpretation of time. This sense of familiarity, despite the lack of any verifiable past experience, suggests the possibility of temporal anomalies, deviations from the strict linear progression of time. Whether these anomalies result from neurological glitches, subconscious processing, or a more fundamental restructuring of our perception of time, remains a subject of ongoing debate and investigation.

Moving beyond subjective experience and anecdotal evidence, we can examine the implications of quantum mechanics for our understanding of time. The non-deterministic nature of quantum phenomena, where particles exist in a superposition of states until measured, directly challenges the concept of a fixed, pre-determined temporal sequence. The very act of measurement influences the outcome, implying an interaction between the observer and the observed that blurs the sharp lines of cause and effect.

Quantum entanglement further complicates this picture. Entangled particles remain interconnected regardless of spatial separation, suggesting a non-local connection that transcends the limitations of space and time. The instantaneous correlation between entangled particles implies a form of temporal interconnectedness that defies the linear model's constraints. If information can be transmitted instantaneously across vast distances, then the notion of a strictly ordered temporal sequence becomes questionable.

Furthermore, the concept of time itself remains a matter of ongoing debate among physicists and philosophers. Is time a fundamental aspect of reality, an independent dimension alongside space, or is it an emergent property arising from the interactions of matter and energy? Different theoretical frameworks offer divergent perspectives, with some suggesting multiple time dimensions or even the possibility of time travel. While these concepts remain largely speculative, they highlight the limitations of a simplistic linear model and point towards a more complex, potentially non-linear reality.

Even in the realm of cosmology, the standard model of the Big Bang suggests a universe that has a beginning and, potentially, an end. But this narrative of a linear cosmological evolution might be overly simplistic. Some cosmologists speculate about cyclical universes, where the Big Bang is not the absolute beginning but rather one stage in a repeating cycle of expansion and contraction. This cyclical model implies a non-linear temporal structure, where the future can influence the past through the cyclical repetition of cosmic epochs.

The implications of shifting from a linear to a non-linear model of time are profound. A non-linear model might better accommodate the complexities of consciousness, memory, subjective experience, and the apparent anomalies observed in parapsychological phenomena. It might also offer a more nuanced understanding of causality, potentially resolving some of the paradoxes associated with time travel or precognition. Such a model might involve a multitude of interconnected temporal threads, creating a richer, more intricate tapestry of time than the simplistic linear model allows.

However, shifting to a non-linear model of time also presents challenges. Our cognitive apparatus is deeply adapted to the linear model; it's how we navigate our world, plan our actions, and understand our place within the universe. A non-linear model demands a significant reframing of our fundamental assumptions about the nature of reality. It requires a willingness to embrace ambiguity, uncertainty, and a more nuanced understanding of cause and effect.

The transition to a more holistic understanding of time necessitates a move away from rigid, reductionist approaches. We must strive towards interdisciplinary collaborations, integrating insights from physics, philosophy, cognitive science, and even parapsychology to develop a more comprehensive model. This model should accommodate both the objective, measurable aspects of time as revealed by physics and the subjective, experiential aspects of time as reflected in our consciousness. The resulting picture may be complex and potentially challenging to our current understanding, but it promises a richer and more complete account of time's nature and its impact on our reality. The journey of embracing non-linearity is an ongoing process of exploration and discovery, leading us towards a deeper understanding of the very fabric of existence. The rewards of this journey – a more profound understanding of ourselves, our universe, and our place within it – are worth the effort. The journey itself is as important as the destination, a continuous process of challenging assumptions, questioning paradigms, and embracing the fascinating mysteries that lie at the heart of time itself.

Overcoming ingrained linear thinking requires a conscious and sustained effort. It's not about abandoning logic or rational thought, but rather augmenting these with a more nuanced understanding of reality. Our minds are wired, through years of education and societal conditioning, to seek linear cause-and-effect relationships. We strive for predictable outcomes, and often find ourselves frustrated when events unfold in unexpected ways. This inherent bias toward linearity can limit our adaptability, creativity, and overall well-being.

One effective strategy is to cultivate mindfulness. Mindfulness practices, such as meditation and deep breathing exercises, help us to quiet the incessant chatter of the mind and become more aware of the present moment. In the present moment, the illusion of linear progression often dissolves. The future, with its projected anxieties and expectations, loses its overwhelming grip, and the past, with its regrets and judgments, fades into the background. We become more receptive to the subtle nuances of our experience, recognizing the interconnectedness of seemingly disparate events.

For example, consider a seemingly minor event—a chance encounter with a stranger. From a linear perspective, this might be dismissed as insignificant. However, a mindful, non-linear perspective allows us to consider the potential ripple effect of this encounter. The conversation might lead to a new opportunity, a fresh idea, or a shift in perspective that fundamentally alters the course of our lives. The point is not to predict these outcomes, but to appreciate the inherent potential within each moment, regardless of how seemingly inconsequential it may appear.

Another powerful technique involves actively challenging our assumptions about causality. We often fall into the trap of assuming a direct, linear relationship between cause and effect. However, reality is rarely so simple. Multiple factors often contribute to any given outcome, and the connections between them are often complex and indirect. Actively questioning our assumptions about cause and effect—seeking out alternative explanations, considering multiple perspectives, and acknowledging the role of chance and randomness—can significantly enhance our ability to think non-linearly.

Consider a business venture that fails. A linear perspective might lead to self-blame or a sense of failure. However, a non-linear perspective would acknowledge the multitude of factors that contributed to the outcome: market conditions, unforeseen competition, internal challenges, and even seemingly random events. Analyzing these factors, rather than focusing solely on a perceived single cause, fosters a more nuanced understanding of the situation and allows for valuable lessons to be learned, leading to greater resilience and adaptability in future endeavors.

Journaling can be a valuable tool in this process. Regularly recording our thoughts, feelings, and experiences allows us to track patterns and connections that might not be readily apparent. By reflecting on past events through a non-linear lens, we can identify subtle influences and unexpected connections, gradually shifting our perception of causality. The act of journaling itself helps to slow down our thought processes, fostering a greater awareness of the present moment and the subtle interplay of events.

Furthermore, engaging in creative activities can be incredibly effective in breaking down linear thinking patterns. Creative endeavors, such as painting, writing, music, or even simply engaging in a playful brainstorming session, encourage exploration, experimentation, and the acceptance of unexpected outcomes. The very nature of creative processes often involves making connections between seemingly unrelated ideas, fostering a more fluid and adaptable approach to problem-solving. The creative process inherently embraces the unexpected, mirroring the non-linear nature of reality.

Cognitive biases also play a significant role in our linear thinking. Confirmation bias, for example, leads us to selectively seek information that confirms our pre-existing beliefs, reinforcing linear patterns of thought. Actively challenging our biases, seeking out alternative viewpoints, and consciously seeking to disprove our own assumptions, are crucial steps in overcoming linear thinking. This requires a commitment to intellectual humility, recognizing the limitations of our own perspectives and the possibility of being wrong.

Similarly, the availability heuristic, which leads us to overestimate the likelihood of events that are easily recalled, can distort our perception of probability and reinforce linear patterns. By consciously acknowledging the limitations of our memory and the potential for bias in our recollections, we can gain a more balanced perspective, leading to a more nuanced understanding of probabilities and a greater capacity for non-linear thinking.

The practice of "lateral thinking," developed by Edward de Bono, offers a structured approach to breaking free from linear thought patterns. Lateral thinking emphasizes exploring alternative perspectives, challenging assumptions, and generating novel solutions by considering seemingly unrelated ideas. Techniques such as "provocation" – deliberately challenging established assumptions – and "random stimulation" – using random inputs to spark new ideas – can help to break down ingrained patterns of linear thinking and foster greater creativity.

Incorporating these techniques into our daily lives requires patience and persistence. Overcoming deeply ingrained mental habits is not a quick fix, but a gradual process of conscious self-awareness and continuous practice. The benefits, however, extend far beyond the intellectual realm. By cultivating a non-linear mindset, we enhance our adaptability, resilience, creativity, and overall well-being. We become better equipped to navigate the complexities of life, appreciate the interconnectedness of events, and embrace the unexpected turns that shape our journeys. The journey towards non-linear thinking is not a destination, but an ongoing process of discovery, a continuous exploration of the rich and multifaceted tapestry of reality. It is a journey worth undertaking.

The shift from a linear to a non-linear perspective extends beyond mere intellectual exercise; it profoundly impacts our psychological and spiritual well-being. The linear model, with its rigid cause-and-effect structure and predictable timeline, inherently fosters a sense of control and predictability. While this can be comforting in the short term, it can also lead to rigidity and an inability to adapt to the unexpected turns life inevitably throws our way. This rigidity manifests as stress, anxiety, and a feeling of being overwhelmed when events deviate from our carefully constructed plans. A non-linear perspective, however, offers a powerful antidote to this stress.

By accepting the inherent fluidity and interconnectedness of events, we reduce the pressure to rigidly control outcomes. Instead of focusing on a singular, predetermined path, we open ourselves to the possibility of multiple pathways and unexpected opportunities. This shift in perspective allows for a greater acceptance of uncertainty, reducing the anxiety associated with the unknown. When we relinquish the illusion of control, we paradoxically gain a sense of freedom and empowerment. We become more adaptable, more resilient, and better equipped to navigate life's challenges with grace and equanimity.

The benefits extend to our creative potential. Linear thinking, by its very nature, confines us within established frameworks and pre-conceived notions. It encourages a convergent approach to problem-solving, focusing on finding the single "right" answer. In contrast, a non-linear perspective fosters divergent thinking, allowing us to explore multiple possibilities and connect seemingly unrelated ideas. This opens up a vast landscape of potential solutions and innovative approaches that would remain hidden under the constraints of linear thought.

Consider the process of artistic creation. A painter working within a strictly linear framework might focus solely on replicating reality with photorealistic precision. However, a painter embracing non-linearity might explore abstract expressionism, allowing the unconscious mind to guide the brushstrokes, leading to unexpected and profoundly moving works of art. Similarly, a composer might break free from traditional harmonic structures, exploring atonal or experimental forms, creating music that transcends conventional expectations and resonates on a deeper level. The non-linear approach to creativity allows for the emergence of originality, innovation, and profound self-expression.

Furthermore, the adoption of a non-linear perspective facilitates personal growth and self-discovery. Linear thinking often leads to rigid self-perception. We become fixated on a predetermined self-image, striving to conform to expectations and resisting change. A non-linear view, however, allows for a more fluid and evolving sense of self. We recognize the inherent dynamism of our personalities, acknowledging the possibility of transformation and growth. We become more open to new experiences, allowing ourselves to adapt and evolve in response to life's challenges and opportunities.

This process of self-discovery is often intertwined with spiritual growth. Many spiritual traditions emphasize the interconnectedness of all things, the illusion of separation, and the cyclical nature of time. These concepts resonate deeply with a non-linear perspective. By recognizing the interconnectedness of events, we begin to appreciate our place within a larger cosmic context. This can lead to a profound sense of meaning, purpose, and belonging, fostering a deep sense of peace and fulfillment.

Consider the concept of karma, a fundamental principle in many Eastern religions. Karma suggests that our actions have consequences that extend beyond immediate cause and effect, rippling through time and space, influencing future events in ways that are not always immediately apparent.

This is a quintessentially non-linear concept, highlighting the complex and interconnected nature of causality. Understanding karma from a non-linear perspective fosters a sense of responsibility and encourages us to act with greater mindfulness and compassion.

Similarly, the cyclical nature of time, often emphasized in various spiritual traditions, aligns with a non-linear worldview. Instead of viewing time as a unidirectional progression from past to future, a cyclical model sees time as a continuous loop, with recurring patterns and cycles of birth, death, and rebirth. This perspective helps us to accept the impermanence of things, reducing our attachment to outcomes and fostering a greater appreciation for the present moment.

The practice of mindfulness, discussed earlier, is a crucial tool in cultivating a non-linear perspective. By focusing on the present moment, we detach from the linear narrative of the past and future, allowing us to perceive the subtle interconnectedness of events. This heightened awareness enhances our intuition, allowing us to recognize patterns and connections that might otherwise remain hidden. This intuitive understanding guides us towards more effective decision-making, leading to greater success and fulfillment in all aspects of life.

The integration of non-linear thinking into our daily lives is not a passive process; it requires conscious effort and practice. It is a journey of self-discovery, a continuous exploration of our perceptions and assumptions. The rewards, however, are immeasurable. By embracing the fluidity and interconnectedness of reality, we reduce stress, enhance our creativity, foster personal growth, and deepen our spiritual understanding. We move from a limited, constricted view of the world to one that is expansive, adaptable, and ultimately, more fulfilling. The journey toward non-linear thinking is a path towards a richer, more meaningful life, a life lived in harmony with the inherent complexity and beauty of existence itself. The more we practice this way of seeing, the more we open ourselves to the infinite possibilities that lie beyond the confines of linear thought. The world unfolds differently when we are ready to see it differently. This is not merely a mental exercise, but a transformative journey of self-discovery and empowerment. It's a path to a more resilient, creative, and fulfilling life. It is, in essence, a path to liberation.

Having explored the multifaceted implications of non-linearity across various domains – from the intricacies of quantum physics to the complexities of human consciousness – we arrive at a pivotal juncture: a re-evaluation of our fundamental understanding of reality. The linear model, deeply ingrained in our Western worldview, has served as a powerful framework for understanding the physical world, enabling remarkable scientific and technological advancements. However, its limitations become increasingly apparent when confronted with the unpredictable nature of reality, particularly at the quantum level and within the realm of human experience.

The journey through this book has unveiled compelling evidence suggesting that reality is far more nuanced, dynamic, and interconnected than the linear model allows. The seemingly random events of quantum mechanics, the intricate feedback loops of complex systems, and the unpredictable nature of human consciousness all point towards a reality that operates on principles beyond simple cause-and-effect relationships. The universe, it seems, is not a clockwork mechanism ticking along a predetermined path, but rather a vibrant, self-organizing system characterized by emergence, feedback, and constant change.

This non-linear perspective necessitates a shift in our understanding of time. The linear conception of time, as a unidirectional flow from past to future, is a simplification, a useful construct for navigating everyday life but ultimately inadequate for grasping the complexities of reality. The non-linear view recognizes the cyclical, iterative, and even multi-dimensional nature of time, drawing parallels with the cyclical patterns found in nature, such as the seasons, biological rhythms, and the ebb and flow of energy. This cyclical view acknowledges the interconnectedness of past, present, and future, emphasizing the influence of past events on the present and the impact of present choices on the future.

This understanding resonates deeply with many spiritual traditions that emphasize the cyclical nature of time and the interconnectedness of all things. Concepts like karma, reincarnation, and the cyclical nature of existence found in various Eastern philosophies align with the non-linear perspective, highlighting the far-reaching consequences of our actions and the cyclical patterns of birth, death, and rebirth. These spiritual frameworks offer valuable insights into the interconnectedness of reality and provide practical tools for navigating life's complexities.

The implications of this revised understanding of reality extend far beyond philosophical contemplation. It fundamentally alters our perception of ourselves and our place in the cosmos. A linear worldview often fosters a sense of separation, a belief in an independent self detached from the larger universe. However, a non-linear perspective reveals the interconnectedness of all things, recognizing our intrinsic connection to the world around us. This realization can lead to a profound sense of awe, wonder, and responsibility, fostering a greater appreciation for the interconnected web of life.

Furthermore, embracing non-linearity has profound implications for our daily lives. It challenges our rigid, often self-limiting beliefs and encourages a more adaptable and resilient approach to life's challenges. The linear model promotes a sense of control, leading to stress and anxiety when events deviate from our carefully constructed plans. A non-linear perspective, however, embraces the unexpected, accepting uncertainty as an integral part of life's tapestry. This shift in perspective allows us to navigate life's complexities with greater grace and equanimity.

Consider the realm of problem-solving. Linear thinking often leads to a convergent approach, focusing on finding a single "right" answer. In contrast, a non-linear perspective fosters divergent thinking, exploring multiple possibilities and unconventional solutions. This approach unlocks creativity and innovation, allowing us to approach problems with greater flexibility and insight. The ability to see connections between seemingly disparate ideas, a hallmark of non-linear thinking, is crucial for breaking free from conventional limitations and developing truly novel solutions.

The impact of this paradigm shift extends to personal growth and self-discovery. Linear thinking often leads to rigid self-perception, fostering a fixed self-image that resists change. A non-linear view, however, recognizes the dynamic and evolving nature of the self, embracing transformation and growth as integral aspects of the human experience. This perspective facilitates greater self-acceptance, allowing us to embrace our imperfections and pursue personal development with greater openness and acceptance.

The integration of non-linear thinking into our lives is not a passive process, but rather an active, ongoing engagement. It requires a willingness to question our assumptions, to challenge our deeply ingrained beliefs, and to embrace the inherent ambiguity of reality. It involves cultivating mindfulness, a practice that allows us to become more aware of our thoughts, feelings, and perceptions, fostering a deeper connection with the present moment and enhancing our ability to perceive the subtle interconnectedness of events. Through mindfulness, we can learn to recognize patterns and connections that might otherwise remain hidden, sharpening our intuition and enabling more effective decision-making.

The journey towards embracing non-linearity is a transformative process, a journey of self-discovery and empowerment. It is a path towards a more resilient, creative, and fulfilling life, one lived in harmony with the inherent complexity and beauty of existence itself. By abandoning the rigid constraints of linear thinking, we open ourselves to the infinite possibilities that lie beyond the confines of conventional understanding. We become more adaptable, more creative, and more attuned to the subtle interconnectedness of all things. This is not simply a shift in perspective; it is a fundamental alteration in our understanding of reality, one that profoundly impacts our relationship with ourselves, with others, and with the universe as a whole.

The exploration of non-linearity presented in this book is not intended to provide definitive answers, but rather to stimulate critical thinking and inspire further inquiry. The insights presented here represent a starting point, a foundation upon which to build a more comprehensive and nuanced understanding of reality. The ongoing dialogue between science and spirituality, between objective observation and subjective experience, is essential for advancing our understanding of the universe and our place within it. By embracing this dynamic interplay, we can move beyond the limitations of the linear model and unlock a deeper, richer appreciation for the extraordinary complexity and interconnectedness of existence. The true understanding of reality may lie not in definitive answers, but in the ongoing process of inquiry, exploration, and acceptance of the inherent uncertainty that characterizes the universe. The journey, it seems, is as important, if not more so, than the destination. The more we engage with this non-linear perspective, the more the universe reveals its mysteries, not in clear pronouncements, but in subtle patterns, emergent properties, and the ever-present dance of interconnectedness. This understanding, ultimately, is not simply intellectual but profoundly transformative, leading to a more fulfilling and harmonious existence. This is not just about grasping a new scientific model; it is about undergoing a profound shift in consciousness, a re-orientation of our being in the world. It is a path toward a deeper, more authentic, and ultimately more meaningful life. The journey, as we have seen, involves not only intellectual understanding but also practical application, the cultivation of mindfulness, and a willingness to embrace the unpredictable nature of reality. This journey is not for the faint of heart, but for those who seek a richer, more profound understanding of themselves and the universe, it is a path of unparalleled richness and reward. The conclusion, therefore, is not an endpoint but a beginning, a call to further exploration, a challenge to embrace the non-linear and experience the boundless possibilities that unfold when we allow ourselves to see the world anew.

15 Further Exploration and Resources

This exploration of non-linearity and its implications across various scientific and spiritual domains has hopefully sparked a thirst for deeper understanding. To that end, this reading list provides a curated selection of resources that delve further into the fascinating concepts we've discussed. This is by no means an exhaustive list, but rather a starting point for your own journey of exploration. The selections are categorized for easier navigation, spanning the fields of physics, philosophy, consciousness studies, and spiritual traditions.

I. Quantum Physics and Non-Linearity:

We began our journey exploring the implications of quantum mechanics, a realm where linearity breaks down and the observer plays an undeniable role. Several key texts offer a deeper dive into this fascinating area:

"Quantum Mechanics: The Theoretical Minimum" by Leonard Susskind and Art Friedman: This accessible introduction to quantum mechanics provides a solid foundation for understanding the fundamental principles that challenge classical, linear perspectives. It avoids overly complex mathematical formalism, focusing instead on the conceptual framework. It's an excellent starting point for anyone with a basic scientific background.

"The Fabric of the Cosmos" by Brian Greene: Greene's work masterfully explains complex concepts in string theory and cosmology, touching upon the challenges to linear causality inherent in these advanced theories. He explores the implications of quantum mechanics for our understanding of space, time, and the universe.

"Quantum Enigma: Physics Encounters Consciousness" by Bruce Rosenblum and Fred Kuttner: This book delves into the often-overlooked role of consciousness in quantum mechanics. It presents the implications of the measurement problem and the challenges this presents to a purely objective, linear view of reality. It is a thought-provoking exploration of the intersection of physics and consciousness.

"Taking the Quantum Leap: The New Physics for Nonscientists" by Fred Alan Wolf: A highly accessible account of quantum physics, this book tackles complex concepts with clear language and insightful analogies. Wolf also touches on the philosophical and spiritual implications of quantum theory, making it a valuable resource for bridging the gap between science and spirituality.

II. Chaos Theory and Complex Systems:

The unpredictable nature of complex systems, as highlighted by chaos theory, provides another perspective on non-linearity. These resources offer deeper insights:

“Chaos: Making a New Science” by James Gleick: Gleick's masterful work provides a comprehensive and engaging overview of chaos theory and its applications across diverse fields. It explains complex concepts in a clear and accessible way, illustrating the importance of non-linearity in understanding natural phenomena.

“Sync: How Order Emerges from Chaos in the Universe, Nature, and Daily Life” by Steven Strogatz: This book explores the fascinating phenomenon of synchronization in complex systems, from fireflies flashing in unison to the coordinated firing of neurons in the brain. It illustrates the spontaneous emergence of order from apparent chaos.

“Introduction to Nonlinear Dynamics” by Steven H. Strogatz: While more mathematically rigorous than Gleick's book, Strogatz's "Introduction to Nonlinear Dynamics" provides a more in-depth exploration of the underlying mathematical principles of chaos theory.

III. Philosophy of Time and Consciousness:

The linear conception of time is a deeply ingrained aspect of our worldview. These philosophical explorations challenge this perspective:

“Time's Arrow and Archimedes' Point” by Huw Price: Price challenges the common-sense view of time's directionality, arguing for a more symmetric view of the past and future. His work offers a profound philosophical critique of our understanding of time.

“The Order of Time” by Carlo Rovelli: Rovelli, a prominent physicist, offers a compelling philosophical exploration of time, blending physics and philosophy to challenge our intuitive notions of time's linearity.

“The Conscious Mind: In Search of a Fundamental Theory” by David Chalmers: Chalmers' seminal work addresses the “hard problem” of consciousness, exploring the relationship between physical processes and subjective experience. This is essential reading for anyone interested in the interaction between mind and matter, and how that might impact our understanding of time.

"The Ego Tunnel: The Science of the Self and the Problem of Consciousness" by Thomas Metzinger: This book explores the nature of the self and the implications of this for our understanding of consciousness and experience, hinting at how our subjective experience of time might be shaped by our cognitive architecture.

IV. Spiritual Traditions and Cyclical Time:

Many spiritual traditions incorporate cyclical conceptions of time, offering insights compatible with the non-linear perspective:

"The Power of Now" by Eckhart Tolle: Tolle's work emphasizes the importance of present moment awareness, challenging the linear preoccupation with past and future. This is a practical guide to living in accordance with a non-linear view of time.

"Tao Te Ching" by Lao Tzu: This ancient Chinese text emphasizes the cyclical nature of reality, the interconnectedness of all things, and the importance of living in harmony with the natural flow of life.

"The Bhagavad Gita": This ancient Hindu scripture, part of the epic Mahabharata, offers profound insights into the nature of time, karma, reincarnation, and the interconnectedness of all beings. It provides a rich spiritual framework for understanding cyclical time.

V. Scientific Papers and Articles:

For those seeking a more academically rigorous approach, exploring relevant scientific papers is highly recommended. Search academic databases like PubMed, JSTOR, and Google Scholar for articles on the following topics:

Non-linear dynamics: Search for papers on specific non-linear systems, such as coupled oscillators, chaotic systems, and complex networks.

Quantum entanglement: Investigate research on the phenomenon of quantum entanglement and its implications for causality and non-locality.

Nonlinear time series analysis: Explore techniques used to analyze data from complex systems that display non-linear behavior.

Consciousness and quantum physics: Look for research exploring the potential connections between consciousness and quantum phenomena.

Mindfulness and brain plasticity: Research on the impact of mindfulness practices on brain structure and function can provide insights into the neural correlates of non-linear thinking.

This recommended reading list offers a diverse range of resources to further explore the concepts discussed in this book. Remember that engaging with these diverse perspectives will require critical thinking, a willingness to question established paradigms, and a commitment to ongoing inquiry. The journey toward a deeper understanding of non-linearity is a continuous process of learning, reflection, and integration. The path forward involves not just passive consumption of information, but active engagement and critical appraisal of the different viewpoints and approaches presented. This is a journey of self-discovery as much as it is a pursuit of scientific and philosophical understanding. Embrace the complexity, engage with the diverse viewpoints, and let the journey itself guide your exploration into the multifaceted nature of reality. The most important aspect is the willingness to question, to explore, and to evolve your understanding of the universe and your place within it. This is an ongoing dialogue, and your continued inquiry is the most valuable contribution.

Rewrite Describe Expand Visualize

Heading 1Heading 2Heading 3Heading 4Heading 5Heading 6Normal Text

Adobe Caslon ProSans-serifArialAustralian SunsetBemboCentury SchoolbookCourierFranklin Gothic MediumFuturaGaramondGaramond Premier ProGeorgiaItc BaskervilleJansonLoraMinion ProPalatinoPapyrusTimes New Roman

8 9 10 11 12 14 18 20 24 30 36 48 60 72 96

1.01.21.51.61.82.02.42.83.04.05.0

Beyond the curated list of books and academic papers, the digital landscape offers a wealth of resources for those eager to delve deeper into the fascinating world of non-linearity, its implications for our understanding of time, consciousness, and the universe. The internet provides access to a diverse range of perspectives, fostering a global conversation on these complex topics. While caution is always warranted when navigating online information, a discerning approach can uncover invaluable insights and connect you with like-minded individuals.

Online Forums and Communities: Several online forums and communities are dedicated to exploring the intersection of science, spirituality, and non-linear thinking. These platforms offer a space for discussion, debate, and the sharing of insights from diverse perspectives. Sites like Reddit (specifically subreddits focused on physics, philosophy, consciousness studies, and alternative science) often host lively discussions on these topics, allowing you to engage with others who share your interests. Be aware, however, that the quality of information varies widely on such platforms; always approach discussions with a critical and questioning mind. Look for communities that encourage rigorous discussion and evidence-based arguments, rather than those prone to unsubstantiated claims and dogmatic pronouncements.

It's crucial to remember that online forums can be echo chambers. Actively seek out diverse viewpoints, even those that challenge your own understanding. This is crucial for intellectual growth and the development of a balanced perspective. Furthermore, be mindful of the potential for misinformation and biased information. Always cross-reference claims with reputable sources and engage in critical thinking.

Websites and Blogs: Numerous websites and blogs are dedicated to exploring non-linearity and its implications. Some focus on specific areas like quantum physics, chaos theory, or consciousness studies, while others offer a broader interdisciplinary approach. Many physics departments and research institutions maintain websites with accessible information on current research. These can provide up-to-date information on the latest advancements in our understanding of non-linear systems. Explore the websites of universities known for their strong research programs in these areas. Often, these universities provide publicly accessible summaries of research findings, making complex information more understandable for a general audience.

Blogs written by experts in the field can offer valuable insights, and many are kept updated with relevant information on scientific advancements. Look for blogs that cite their sources and engage in transparent discussion of methodology. Be aware, however, that some blogs may promote particular viewpoints or theories without sufficient evidence. It is vital to maintain a critical approach and to evaluate the credibility of the information presented.

Podcasts and Online Lectures: The world of audio and video offers a wealth of resources to further enhance your understanding. Many podcasts dedicate episodes or entire series to the exploration of non-linearity and related concepts. These podcasts often feature interviews with leading researchers and thinkers, providing diverse perspectives and insights. These are often highly accessible and engaging ways to learn about complex topics. Search for podcasts focusing on science, philosophy, consciousness, or spirituality to find relevant content. Pay attention to the credibility of the guests and the rigor of the discussion. Always verify claims made on podcasts with other sources.

Similarly, online lectures and video courses from universities and educational platforms (like Coursera, edX, and YouTube channels from academic institutions) often provide excellent educational resources on relevant topics, such as quantum mechanics, chaos theory, and cognitive science. These can offer in-depth explanations of complex concepts, complementing the more general introductory materials found elsewhere. These lectures frequently offer a more structured and comprehensive approach to the topics, often including exercises and assignments to deepen the understanding.

Specialized Databases and Academic Journals: For those seeking the most rigorous and up-to-date information, academic databases like JSTOR, ScienceDirect, and PubMed offer access to a vast collection of peer-reviewed research papers and articles. These databases are treasure troves of information but require some familiarity with scientific literature. These resources are invaluable for those with more advanced knowledge of the topics; using these effectively requires some degree of familiarity with scientific terminology and research methods.

Focus your searches on keywords such as "non-linear dynamics," "complex systems," "quantum entanglement," "consciousness studies," "time perception," and "cognitive science." Remember that navigating this landscape can take time and effort; it's wise to start with review papers that summarize the current state of knowledge within specific areas. This approach helps you gain an overview of the research before delving into highly technical specialized papers.

Utilizing Online Resources Effectively: While online resources offer incredible potential, navigating them effectively requires a critical and discerning approach. Always cross-reference information from multiple reputable sources and be wary of websites or individuals promoting unsubstantiated claims or biased perspectives. Remember that the internet hosts a wide spectrum of perspectives, from rigorously researched and evidence-based claims to outright misinformation. Engage with the information critically, utilizing your own judgment and skepticism to determine the validity of the material you encounter.

The Importance of Community and Critical Engagement: The study of non-linearity and its implications is an ongoing exploration. By engaging with online communities, participating in discussions, and critically evaluating the information you encounter, you contribute to a global conversation about some of the most profound questions facing humanity. This ongoing engagement is vital, both for deepening your own understanding and for advancing our collective knowledge of the universe and our place within it. Remember that the pursuit of knowledge is a continuous process, and your active engagement is a crucial aspect of that journey. The combination of individual study and community interaction will significantly enhance your understanding and appreciation of non-linear systems. Embrace this dynamic and evolving learning experience, and allow your critical inquiry to guide your exploration.

The exploration of non-linearity, as we've journeyed through this book, has unveiled a fascinating and complex landscape. While significant progress has been made in understanding various aspects of non-linear systems, numerous questions remain unanswered, beckoning further investigation. These unresolved issues span multiple disciplines, highlighting the inherently interdisciplinary nature of this field of inquiry.

One of the most pressing areas requiring further research concerns the precise mechanisms through which non-linearity manifests in consciousness. While we've examined the potential links between non-linear dynamics and subjective experience, the exact nature of this interaction remains elusive. How do non-linear processes in the brain, for example, translate into the qualitative aspects of conscious awareness? Are there specific neural correlates of non-linear temporal perception, and how can we identify and measure them objectively? Sophisticated neuroimaging techniques, combined with advanced mathematical modeling, could provide crucial insights into these questions. Furthermore, exploring the potential role of quantum processes in consciousness, given the inherent non-linearity of quantum mechanics, remains a fertile ground for future investigations. This could involve exploring the potential influence of quantum entanglement on neural processes, or examining the role of quantum coherence in maintaining integrated conscious experience.

Another critical area needing attention is the development of more robust mathematical models and computational tools to simulate and analyze non-linear systems. While existing models such as chaos theory and complex systems theory have provided valuable frameworks, they are often limited in their ability to accurately capture the intricacies of real-world non-linear phenomena. Developing more sophisticated algorithms and computational methods that can handle the high dimensionality and unpredictable nature of these systems is essential. This might involve exploring the application of artificial intelligence and machine learning techniques to analyze large datasets of non-linear temporal data. Such approaches might facilitate the discovery of hidden patterns and relationships that could help us better understand the underlying mechanisms at play. Moreover, developing more powerful computing infrastructure will be essential to handle the computational demands of these advanced models.

The integration of different theoretical frameworks represents another significant challenge. Currently, there is a tendency to compartmentalize the study of non-linearity across various disciplines, such as physics, biology, and cognitive science. However, a more holistic and integrated approach is needed. For example, bridging the gap between the mathematical models of non-linear dynamics and the phenomenological descriptions of subjective experience requires a careful interdisciplinary dialogue. This could involve the development of new theoretical frameworks that can accommodate both the quantitative rigor of mathematical models and the qualitative richness of human experience. Such cross-disciplinary collaboration will prove invaluable in achieving a more unified and comprehensive understanding of non-linear temporal processes.

Furthermore, the ethical implications of understanding non-linearity, particularly its potential application in areas such as predictive modeling and artificial intelligence, warrant careful consideration. As our ability to model and predict the behavior of non-linear systems improves, so does the potential for misuse. For example, highly accurate prediction models could be used for manipulative purposes, infringing on individual autonomy or exacerbating existing social inequalities. Robust ethical guidelines and regulatory frameworks must be developed to ensure that advancements in the understanding of non-linearity are used responsibly and ethically. These ethical considerations will be paramount as technology progresses, and it is imperative that scientists, policymakers, and the public engage in open and transparent dialogues to navigate these challenges.

Another vital area for future research involves investigating the relationship between non-linearity and the concept of free will. If our actions are influenced by non-linear systems that exhibit unpredictable behavior, to what extent do we truly possess free will? This is a philosophical question with profound implications for our understanding of personal responsibility and moral agency. Exploring the interplay between deterministic processes and subjective agency within non-linear frameworks requires interdisciplinary engagement with philosophy, neuroscience, and cognitive science. Understanding the role of consciousness in navigating and potentially influencing non-linear processes will be crucial in resolving this persistent debate.

The study of non-linear temporal perception opens another avenue for investigation. How does our subjective experience of time relate to the underlying non-linear dynamics of our brains and bodies? Do individual differences in temporal perception reflect variations in the non-linear characteristics of these systems? Developing reliable methods for measuring and quantifying temporal perception, along with advanced neuroimaging techniques, might allow researchers to examine these relationships more precisely. Furthermore, investigating the effects of altered states of consciousness (such as meditation or drug-induced states) on temporal perception could shed light on the neural mechanisms underlying the subjective experience of time.

Research into the role of non-linearity in biological systems beyond the nervous system also presents significant opportunities. For instance, exploring the non-linear dynamics of ecological systems, such as predator-prey relationships or climate patterns, could provide insights into the complex interactions within these environments. Understanding the non-linear aspects of such systems is vital for developing effective strategies for conservation and environmental management. Likewise, investigating non-linearity in the realm of genetics and molecular biology could reveal new understandings of genetic regulation, evolution, and disease processes. The non-linear interactions within these biological systems hold the key to understanding how simple initial conditions can lead to profound and unpredictable outcomes.

Finally, the exploration of non-linearity extends into the realm of cosmology and the study of the universe at large. From the chaotic dynamics of planetary systems to the potentially non-linear nature of spacetime itself, the universe exhibits a multitude of non-linear phenomena. Investigating the role of non-linearity in the formation of galaxies, the evolution of stars, and the expansion of the universe could lead to significant advancements in our cosmological understanding. Exploring the relationship between non-linearity and the fundamental forces of nature, particularly gravity, remains a significant challenge, potentially requiring breakthroughs in our understanding of quantum gravity.

In conclusion, the journey into the world of non-linearity has just begun. The open questions and challenges outlined here represent a rich landscape for future research. By embracing interdisciplinary collaboration, developing innovative methodologies, and engaging in critical analysis, we can unlock further insights into the profound implications of non-linearity for our understanding of time, consciousness, and the universe itself. The exploration of these complexities not only promises scientific advancements but also offers the potential for transformative shifts in our philosophical perspectives and our understanding of our place within the cosmos. The continued investigation of non-linear systems holds the key to unlocking deeper mysteries of reality and enriching our understanding of ourselves and the universe we inhabit. The future of this field is bright, filled with the promise of exciting discoveries and potentially paradigm-shifting insights.

Engaging with the multifaceted nature of non-linearity requires a multifaceted approach. This isn't a journey undertaken solely within the confines of a textbook or research paper; it's a journey of exploration that extends far beyond the pages of this book. To deepen your understanding and contribute to this evolving field, I encourage you to consider the following avenues of engagement.

Firstly, consider delving into hands-on experimentation. While many of the examples discussed herein involve complex mathematical modeling and sophisticated technological equipment, opportunities exist for engaging with non-linearity in more accessible ways. For instance, simple pendulum experiments can visually demonstrate the unpredictable nature of chaotic systems. By carefully observing the pendulum's swing under varying initial conditions and levels of friction, one can witness firsthand the sensitive dependence on initial conditions – a hallmark of non-linear systems. Similar experiments can be conducted with double pendulums, showcasing even more complex and unpredictable behavior. The results can be analyzed using basic mathematical tools, offering a concrete experience of non-linear dynamics. Furthermore, the readily available computational tools, such as simulations of the Lorenz attractor or other chaotic systems, can offer another avenue for experimentation and visualization. The beauty of these simulations is that they allow for hands-on manipulation of parameters, directly observing the effects on the system's behavior. This interactive approach can significantly improve comprehension.

Moving beyond the realm of physics, biological systems provide a rich field for experimental exploration. Observing the growth patterns of certain plants or the branching patterns of trees can offer fascinating insights into non-linear biological processes. These patterns often exhibit fractal characteristics, reflecting the self-similarity inherent in many non-linear systems. Similarly, studying the dynamics of population growth in simple ecological systems, either through simulations or observations of real-world populations, can highlight the unpredictable and often counter-intuitive nature of non-linear relationships. While sophisticated lab equipment is often required for the most in-depth biological investigations, simpler observations can be incredibly insightful and pave the way for a deeper appreciation of the principles at play.

Beyond experimental approaches, engaging in philosophical discussions around the implications of non-linearity is crucial. The challenge of reconciling the seemingly deterministic nature of physical laws with the unpredictable behavior of non-linear systems presents a fertile ground for philosophical inquiry. Consider the implications of non-linearity for our understanding of free will, determinism, and causality. Does the inherent unpredictability of non-linear systems challenge the traditional notion of causality? How do we reconcile the seemingly random fluctuations within these systems with our understanding of predictable, cause-and-effect relationships? These questions necessitate engagement with philosophical traditions ranging from classical determinism to contemporary process philosophy. Actively participating in philosophical debates, either through reading relevant literature or engaging in discussions with others, can enrich your comprehension of the broader implications of non-linear systems.

Participating in relevant communities is another vital aspect of engaging further with this topic. Connecting with researchers, scientists, and enthusiasts working in related fields can broaden your perspective and foster a more comprehensive understanding. Attending conferences, workshops, and seminars focused on complexity science, non-linear dynamics, or related disciplines provides opportunities to learn from experts, network with colleagues, and contribute to ongoing discussions. Online forums, discussion groups, and social media communities dedicated to these fields offer valuable platforms for exchanging ideas, asking questions, and staying updated on the latest research. Active participation in these communities fosters collaborative learning and accelerates your understanding. Moreover, contributing to open-source projects related to non-linear dynamics or participating in citizen science initiatives can provide direct involvement in advancing the field.

Furthermore, exploring the intersection of non-linearity and various artistic expressions can reveal unexpected insights. Art, music, and literature often reflect the underlying principles of complex systems and non-linear dynamics. Consider the fractal patterns found in nature, which have inspired numerous artistic creations. The seemingly random yet structured patterns in these artworks echo the unpredictable yet organized behavior characteristic of many non-linear systems. Likewise, the improvisational nature of jazz music, with its emphasis on spontaneity and emergent structures, offers a compelling analogy for the unpredictable yet structured dynamics found in complex systems. Exploring these artistic expressions can reveal the deep connections between the scientific and the aesthetic, fostering a more holistic appreciation of non-linearity's impact.

Specifically, in the context of consciousness studies, the implications of non-linearity are profound and require careful consideration. The exploration of consciousness through the lens of non-linear dynamics necessitates a multidisciplinary approach, integrating insights from neuroscience, physics, philosophy, and even parapsychology. The challenging task is to develop models capable of capturing the richness and complexity of subjective experience within a framework that acknowledges the unpredictable nature of non-linear processes. This requires a willingness to embrace ambiguity and uncertainty while remaining rigorous in the pursuit of scientific understanding. The development of new methodologies that bridge the gap between qualitative descriptions of consciousness and quantitative models of non-linear systems remains a crucial challenge, and one requiring significant interdisciplinary effort.

Exploring the intersection of non-linearity and artificial intelligence presents another fascinating and rapidly evolving area of research. The development of artificial neural networks, inspired by the structure and function of the human brain, is inherently intertwined with the principles of non-linear dynamics. These networks exhibit complex and unpredictable behavior, often exceeding our ability to fully understand their internal processes. This highlights the potential and the limitations of AI systems, raising crucial questions about predictability, control, and the very nature of intelligence itself. Understanding the non-linear dynamics of artificial neural networks is crucial for developing more robust, reliable, and ethically responsible AI systems.

The ethical implications of understanding and applying non-linear dynamics cannot be overstated. As our ability to model and predict the behavior of complex systems increases, so does the potential for both beneficial and harmful applications. The development of sophisticated predictive models based on non-linear dynamics has far-reaching implications across various sectors, from finance and weather forecasting to public health and national security. However, these advancements necessitate careful ethical consideration to prevent misuse or unintended consequences. Robust ethical frameworks and regulatory mechanisms must be developed to guide the responsible application of this knowledge, ensuring that it serves the benefit of humanity while safeguarding against potential risks. Open dialogue involving scientists, policymakers, and the public is essential to navigate the complex ethical landscape presented by advancements in non-linear science.

In conclusion, the exploration of non-linearity is a journey that requires continuous learning, experimentation, and critical engagement. By embracing a multidisciplinary approach, engaging in thoughtful discussions, participating in relevant communities, and remaining mindful of the ethical implications, we can significantly advance our understanding of this profound and ever-evolving field. This ongoing exploration promises not only scientific breakthroughs but also a deeper understanding of ourselves, our place within the universe, and the intricate dance of order and chaos that shapes our reality. The future of non-linear science is brimming with possibilities, and your active participation is vital in shaping its trajectory.

Embracing the unknown is not merely a philosophical stance; it's a pragmatic necessity when navigating the complexities of non-linear systems. The inherent unpredictability of these systems challenges our ingrained desire for certainty and control. We are accustomed to linear models, where cause and effect follow predictable paths. However, the world of non-linearity reveals a different reality – one characterized by emergent behavior, sensitive dependence on initial conditions, and a fundamental limit to our predictive capabilities. Accepting this limitation is not a surrender to ignorance; rather, it's a recognition of the inherent nature of reality and a crucial step toward a deeper understanding.

This acceptance requires a shift in perspective, from a worldview focused on precise prediction to one that embraces the dynamic interplay of order and chaos. It involves relinquishing the comfort of linear models and venturing into the uncharted territory of non-linearity, where patterns emerge from seemingly random interactions and where the whole is often greater than the sum of its parts. This is not a retreat from scientific rigor, but a recalibration of our methodologies. It calls for a more nuanced approach, one that acknowledges the limitations of our current models and embraces the inherent uncertainty of complex systems.

One implication of this uncertainty is the need for a more adaptable and flexible approach to research. Rigid methodologies, designed for linear systems, may prove inadequate when dealing with the unpredictable nature of non-linearity. This suggests the necessity of developing new research methods, perhaps incorporating elements of exploratory research, qualitative analysis, and iterative modeling. The focus should shift from precise prediction to understanding the underlying mechanisms and emergent properties of these systems, even if complete predictability remains elusive. We must develop a scientific intuition attuned to the unexpected, ready to embrace surprising results and adapt our models accordingly.

Furthermore, the acceptance of uncertainty necessitates a reassessment of our understanding of causality. In linear systems, causality is often straightforward: A causes B. However, in non-linear systems, the relationship between cause and effect can be far more intricate and less predictable. Multiple factors may interact in complex ways, leading to emergent outcomes that are not readily apparent from the initial conditions. This challenges the traditional view of causality, requiring a more nuanced understanding that acknowledges the interplay of multiple factors and the possibility of unforeseen consequences. The focus should shift from identifying single, direct causes to understanding the complex interplay of factors contributing to an outcome.

The implications extend beyond the purely scientific realm. Our social and political systems, for example, often exhibit characteristics of non-linearity. Small, seemingly insignificant events can trigger significant and unpredictable changes, highlighting the limitations of simple cause-and-effect models in understanding social dynamics. Understanding this non-linearity is crucial for developing more effective strategies for addressing complex social challenges, ranging from climate change to economic instability. This requires embracing the inherent unpredictability of social systems and developing strategies that are adaptable and resilient in the face of unexpected events.

The journey into the realm of non-linearity also compels a deeper engagement with philosophical questions. The unpredictability of these systems raises profound questions about free will versus determinism, the nature of causality, and even the limits of human knowledge. These are not simply academic questions; they bear upon our understanding of ourselves and our place in the universe. Engaging with these philosophical issues requires a willingness to grapple with ambiguity and a recognition that definitive answers may often be elusive. However, the very process of engaging with these questions can enrich our understanding of the profound implications of non-linearity.

The exploration of non-linear systems extends into the realm of consciousness studies. The human brain, a vastly complex network of interacting neurons, undoubtedly exhibits non-linear dynamics. Understanding how non-linearity manifests in brain activity, and its relation to conscious experience, remains a major challenge for neuroscience and cognitive science. The quest to understand consciousness may require shifting from reductionist approaches to more holistic models that incorporate non-linear principles. This interdisciplinary approach will demand collaboration among neuroscientists, physicists, philosophers, and even those exploring more unconventional approaches, such as parapsychology. The potential breakthroughs from this collaboration could revolutionize our understanding of consciousness and the nature of reality itself.

This exploration extends to artificial intelligence as well. Artificial neural networks, which underly many AI systems, function through non-linear processes. The unpredictable nature of these networks highlights both the potential and limitations of AI. The development of truly intelligent AI requires not only the ability to replicate certain human cognitive functions, but also to grasp the non-linear aspects of consciousness and cognition. This requires a cautious, ethical approach, balancing the potential benefits with the potential risks. Our future interaction with advanced AI will be deeply intertwined with the understanding of non-linearity.

Finally, consider the ethical implications. As our ability to model and predict non-linear systems improves, so does the potential for both beneficial and detrimental applications. Advanced predictive models based on non-linear dynamics will have widespread implications across diverse sectors, from finance and healthcare to environmental management and national security. However, this power comes with responsibility. The development and application of non-linear science demands rigorous ethical considerations to prevent misuse and unintended consequences. Open dialogue among scientists, policymakers, and the public is crucial to establish robust ethical frameworks and ensure that these advances serve humanity's best interests.

In conclusion, our journey through the landscape of non-linearity underscores the importance of embracing uncertainty and the unknown. This isn't a retreat from rational inquiry; it's a recalibration of our approach to understanding the world. It requires a willingness to challenge our assumptions, embrace ambiguity, and develop new methodologies tailored to the unpredictable nature of complex systems. By engaging in interdisciplinary dialogue, actively participating in research, and fostering a culture of ethical responsibility, we can navigate the complexities of non-linearity and unlock its potential for profound scientific and societal advancements. The path forward demands curiosity, adaptability, and a recognition that the most fascinating discoveries often emerge from the most unexpected places. The future of non-linear science is not a destination, but a continuous exploration, and your involvement is vital in shaping its direction.

Citations

The Traditional Linear Perspective of Time

Hawking, Stephen. *A Brief History of Time*. Bantam Books, 1988.

Smolin, Lee. *Time Reborn: From the Crisis in Physics to the Future of the Universe*. Houghton Mifflin Harcourt, 2013.

Time in Human Consciousness and Everyday Experience

Zimbardo, Philip, and John Boyd. *The Time Paradox: The New Psychology of Time That Will Change Your Life*. Free Press, 2008.

Damasio, Antonio. *The Feeling of What Happens: Body and Emotion in the Making of Consciousness*. Harcourt Brace, 1999.

Introduction to Non-Linear Time and Its Implications

Barbour, Julian. *The End of Time: The Next Revolution in Physics*. Oxford University Press, 1999.

Rovelli, Carlo. *The Order of Time*. Riverhead Books, 2018.

Historical Perspectives on Time

Whitrow, G.J. *Time in History: Views of Time from Prehistory to the Present Day*. Oxford University Press, 1988.

Theories Challenging Linear Time

Gödel, Kurt. "A Remark About the Relationship Between Relativity Theory and Idealistic Philosophy." In *Albert Einstein: Philosopher-Scientist*, edited by Paul Arthur Schilpp, Open Court, 1949.

Penrose, Roger. *The Emperor's New Mind: Concerning Computers, Minds, and the Laws of Physics*. Oxford University Press, 1989.

Philosophical Implications of Non-Linear Time

McTaggart, J.M.E. "The Unreality of Time." *Mind*, vol. 17, no. 68, 1908, pp. 457-474.

Merleau-Ponty, Maurice. *Phenomenology of Perception*. Routledge, 1945.

Introduction to the Mandela Effect

Broome, Fiona. *The Mandela Effect: Confabulation or Fact?*, New Consciousness Review, 2009.

Notable Examples and Case Studies

Radin, Dean. *The Conscious Universe: The Scientific Truth of Psychic Phenomena*. HarperOne, 1997.

Psychological and Sociological Explanations

Loftus, Elizabeth. *Eyewitness Testimony*. Harvard University Press, 1996.

Bartlett, Frederic C. *Remembering: A Study in Experimental and Social Psychology*. Cambridge University Press, 1932.

Link Between Non-Linear Time and the Mandela Effect

Bohm, David. *Wholeness and the Implicate Order*. Routledge, 1980.

Quantum Mechanics and Time

Feynman, Richard P., and Albert R. Hibbs. *Quantum Mechanics and Path Integrals*. McGraw-Hill, 1965.

Theories of Time Loops, Jumps, and Parallel Universes

Everett, Hugh. "The Theory of the Universal Wave Function," Princeton University, 1957.

Deutsch, David. *The Fabric of Reality: The Science of Parallel Universes—and Its Implications*. Penguin Books, 1997.

Evidence from Physics and Cosmology

Greene, Brian. *The Fabric of the Cosmos: Space, Time, and the Texture of Reality*. Vintage Books, 2005.

Tegmark, Max. *Our Mathematical Universe: My Quest for the Ultimate Nature of Reality*. Knopf, 2014.

Personal Anecdotes and Stories

Monroe, Robert A. *Journeys Out of the Body*. Doubleday, 1971.

Cultural and Mythological Representations of Non-Linear Time

Eliade, Mircea. *The Myth of the Eternal Return: Cosmos and History*. Princeton University Press, 1954.

The Impact of Non-Linear Time on Identity and Consciousness

Targ, Russell, and Harold Puthoff. *Mind-Reach: Scientists Look at Psychic Abilities*. Delacorte Press, 1977.

The Brain's Perception of Time

Eagleman, David. *Incognito: The Secret Lives of the Brain*. Pantheon, 2011.

Cognitive Biases and Temporal Illusions

Kahneman, Daniel. *Thinking, Fast and Slow*. Farrar, Straus and Giroux, 2011.

How Non-Linear Time Shapes Our Reality

Bohm, David. *Thought as a System*. Routledge, 1992.

Ancient Astronaut Theories and the Anunnaki

Sitchin, Zecharia. *The 12th Planet*. Harper, 1976.

Connections Between Ancient Myths and Modern Non-Linear Time Theories

Hancock, Graham. *Fingerprints of the Gods: The Evidence of Earth's Lost Civilization*. Crown, 1995.

Implications for Our Understanding of History and Civilization

Von Däniken, Erich. *Chariots of the Gods? Unsolved Mysteries of the Past*. Putnam, 1968.

Representation of Non-Linear Time in Literature, Film, and Television

Jameson, Fredric. *Postmodernism, or, the Cultural Logic of Late Capitalism*. Duke University Press, 1991.

Influence on Storytelling and Narrative Structure

McKee, Robert. *Story: Substance, Structure, Style and the Principles of Screenwriting*. HarperCollins, 1997.

Case Studies: "Doctor Who," "Interstellar," "Dark," and More

Nicholls, Peter. *The Encyclopedia of Science Fiction*. St. Martin's Press, 1993.

Weaving Together Science, Philosophy, and Personal Experience

Capra, Fritjof. *The Tao of Physics: An Exploration of the Parallels Between Modern Physics and Eastern Mysticism*. Shambhala Publications, 1975.

The Future of Non-Linear Time Research

Kaku, Michio. *The Future of the Mind: The Scientific Quest to Understand, Enhance, and Empower the Mind*. Doubleday, 2014.

How Embracing Non-Linear Time Can Enrich Our Understanding of Reality

Sheldrake, Rupert. *The Presence of the Past: Morphic Resonance and the Habits of Nature*. Times Books, 1988.

www.ingramcontent.com/pod-product-compliance
Lightning Source LLC
LaVergne TN
LVHW090319210926
848480LV00001B/295
9798230596172